TALK ABOUT A MESS

TALK ABOUT A MESS

It Happened in Vietnam

Terrance R. Dinan

Published by Mess Hall Publishing Group
New York, NY

www.messhallpublishing.com

ISBN: 978-0-9891507-0-5

Printed in the United States of America

For Anne

Kinsey and Ashley

Evelyn and Sebastian

CONTENTS

PREFACE

W*HY Vietnam?* was an informational film starring President and Commander-in-Chief Lyndon Baines Johnson. By his command, it was required viewing for all active military personnel in 1965. As an intelligence analyst with the rank of corporal in the Army of the United States, Fort Holabird, Maryland, I dutifully attended this fateful and fanciful fiction.

Two years later, following completion of the Officer Candidate School and the Open Mess Management Course at Fort Lee, Virginia, I was the only member of my OCS graduating class whose first duty assignment was Vietnam. I was required to obtain a Top Secret clearance before leaving to replace a senior captain in an advisory team in South Vietnam. Secrecy being sacrosanct, no other details regarding my deployment were divulged.

Thus it was that in April 1967, I arrived in South Vietnam without a clue as to what lay ahead. I was twenty-three years old. I had never traveled outside the U.S.A. Now I was assigned to a war zone on the opposite side of the globe. I was more than apprehensive, I was fearful; mostly, I simply hoped to survive. Little did I know that I was embarking on the education of a lifetime.

Over the next many months, I learned about an exceptional culture and its people in a remarkably beautiful and abundant land, even as it was caught in the disaster of war. I learned to understand the qualities of true leadership while experiencing the exercise of power and position at its best and at its worst. I experienced firsthand the imperative of meeting on the level with all our fellow travelers along this transitory plane; travelers who are, after all, leveled by legitimate or imagined fears, hopes, aspirations, and longings.

And under exhausting and trying conditions, I experienced the light of levity. Though dark and daunting as many encounters in life may be, there bubbles to the surface that humorous element inherent in most human endeavors that somehow balances our thinking and keeps our heads about us when reason is unreasonable. I did survive, and I returned home with profound admiration and respect for the people of Vietnam, for the safety and freedom of the United States of America, and for the leadership and integrity of the U.S. military.

If it had somehow occurred to me lo those many years ago that I would someday feel compelled to write this memoir, I would have taken a lot more photographs, and more importantly, I would have recorded the names of everyone I encountered. Though my memories of these people and their actions remain vivid, time has taken it inevitable toll. I can still see them clearly: their faces, their postures, their rank insignia; I just can't read the names on their shirts. Thus it is that although many names herein are accurate, others, by necessity, are my inventions—an author's prerogative, I suppose.

In organizing the stories of these unforgettable happenings, I have followed the timeline of thought association rather than absolute chronology. And so it happens, for example, that the behind-the-scenes maneuvering involved in the convoluted conversion from COLA to Class 1 appears in chapter twenty-three, yet that same conversion is mentioned in chapter twenty as a fait accompli—a memoir's prerogative, I suppose.

Having now learned a bit about navigating the web, I have been amazed and delighted to find that soldiers who ate at the Eakin Compound Messes still remember our waitress service and choice of entrees. One even recalls the humor of finding Yankee pot roast and Southern fried chicken on the same menu.

President Johnson's rhetorical question "Why Vietnam?" may never be suitably answered. The singular reality that stands out above all the happenings, real and imagined, is the almost universal respect and admiration shared between the Vietnamese and the Americans who were brought together during the United States' involvement in Vietnam.

Many a child in every place and circumstance at one time or another asks the universal question, "What was it really like in the war, Daddy?" There are as many true answers as there are survivors. I learned to love and respect the finest aspects of our military, and I gained a particular high regard for Vietnam, its culture, and especially its people. I hope you, the reader, will find laughter, entertainment, and understanding in this book, my personal answer to that innocent question.

East China Sea
CHINA
INDIA
TAIWAN
BURMA
HONG KONG
Gulf of Tonkin
LAOS
Philippine Sea
THAILAND
South China Sea
CAMBODIA
Andaman Sea
PHILIPPINES
Gulf of Thailand
MALAYSIA
Celebes Sea
INDONESIA
Indian Ocean

SOUTHEAST ASIA

NORTH and SOUTH VIETNAM

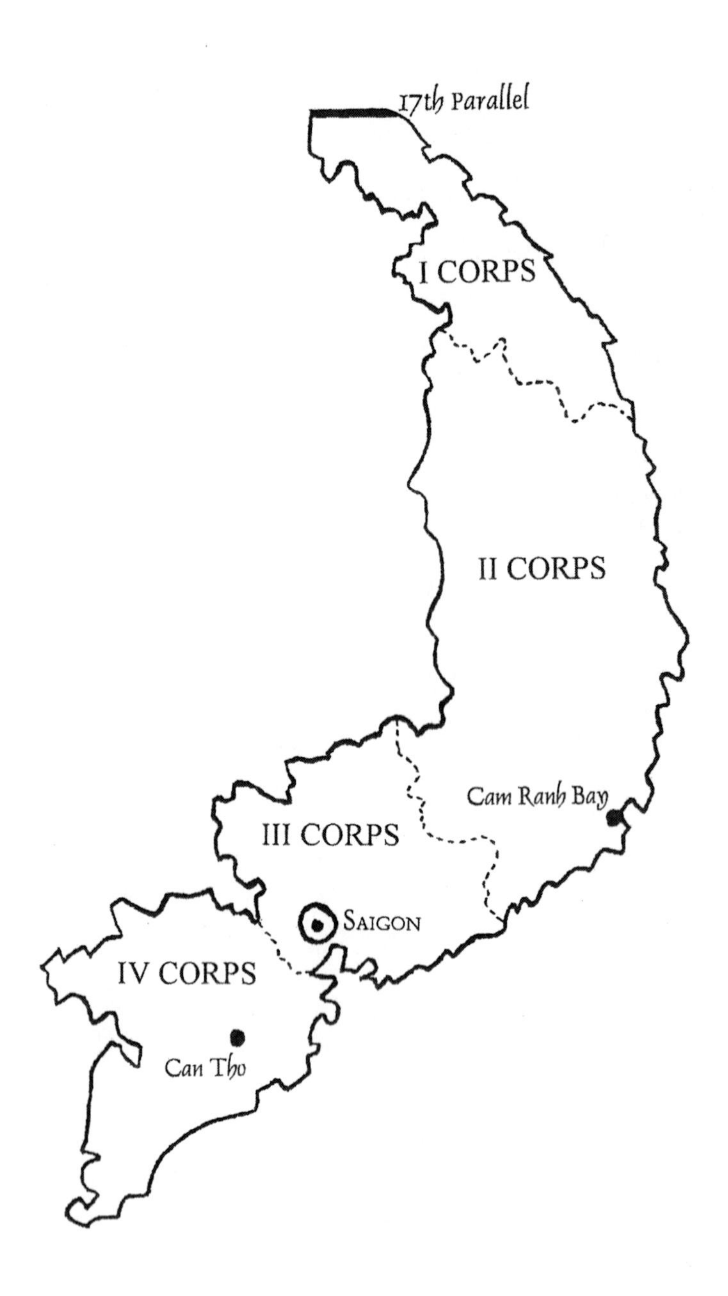

SOUTH VIETNAM'S FOUR MILITARY REGIONS

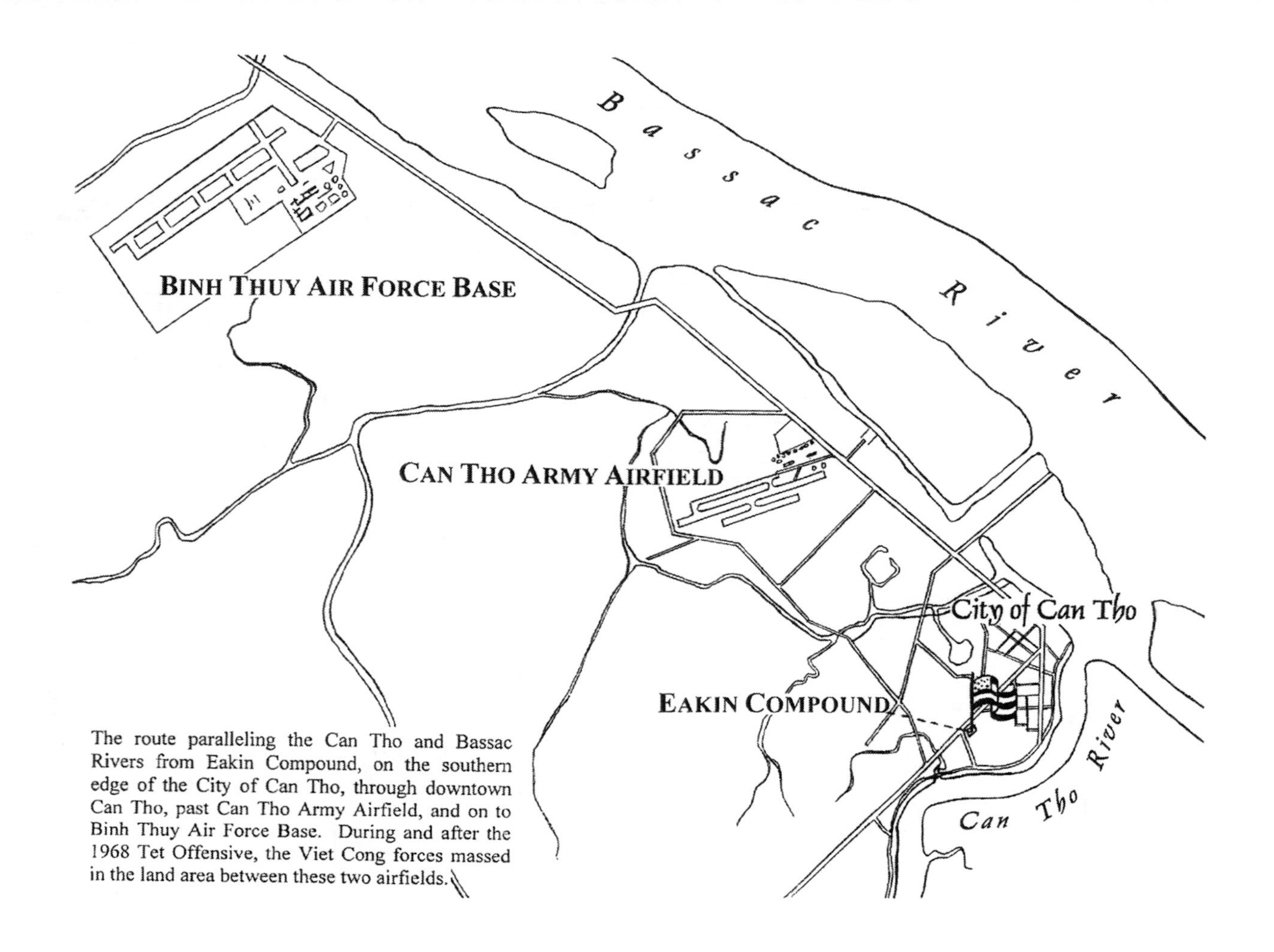

Bassac River
Binh Thuy Air Force Base
Can Tho Army Airfield
City of Can Tho
Eakin Compound
Can Tho River
The route paralleling the Can Tho and Bassac Rivers from Eakin Compound, on the southern edge of the City of Can Tho, through downtown Can Tho, past Can Tho Army Airfield, and on to Binh Thuy Air Force Base. During and after the 1968 Tet Offensive, the Viet Cong forces massed in the land area between these two airfields.

GENERAL AREA MAP

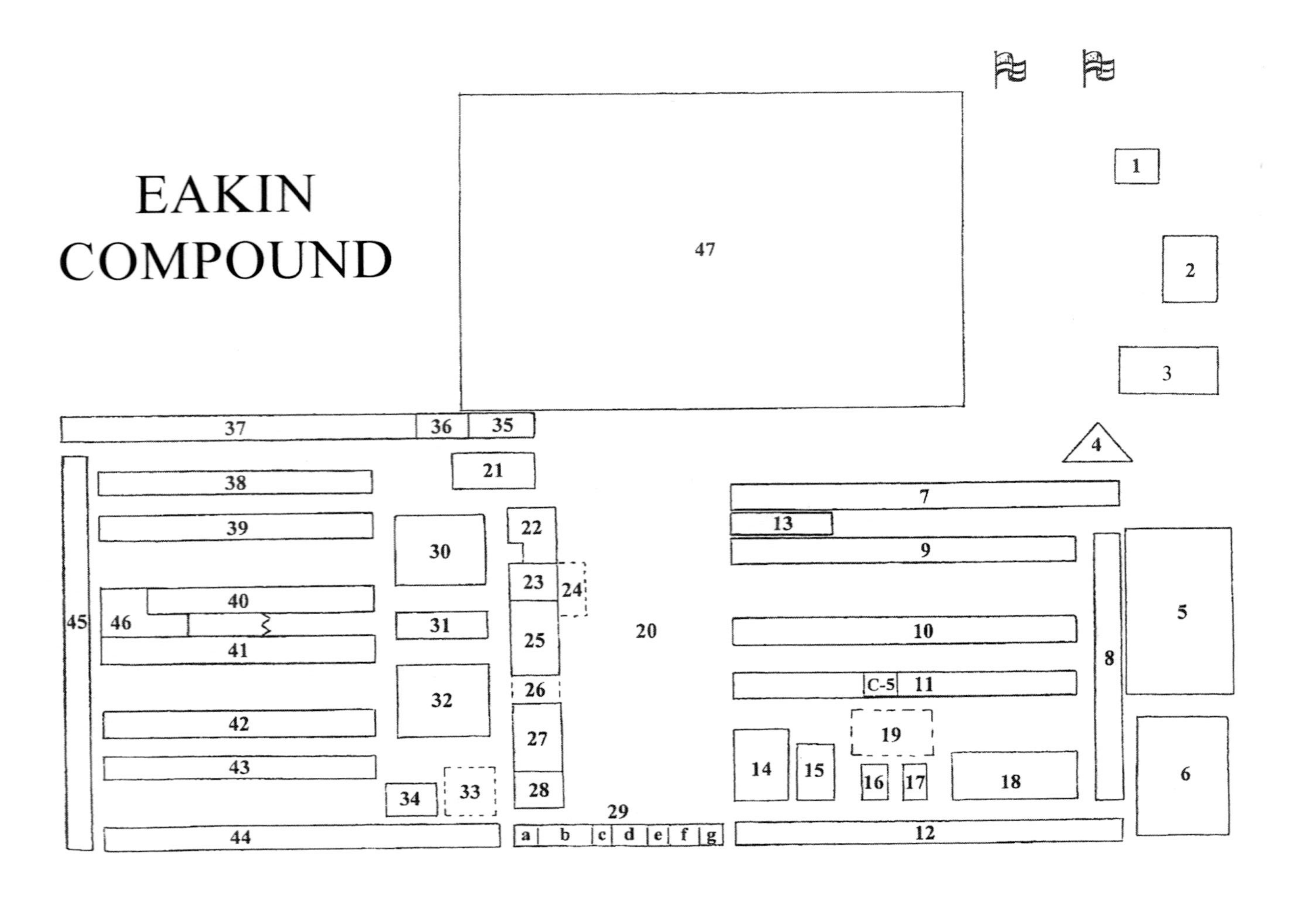
EAKIN
COMPOUND
47
1
2
3
4
37
36
35
21
38
39
30
22
23
24
40
45
46
31
25
20
41
26
32
42
27
43
34
33
28
29
44
a
b
c
d
e
f
g
7
13
9
10
C-5
11
19
14
15
16
17
18
12
8
5
6

Entrance from Can Tho's Main Road

1 Checkpoint for the Vietnamese

2 Warehouse / Grocery

3 Vietnamese Guard Barracks

4 Machine Gun Emplacement

5 Tennis Court

6 Swimming Pool

7 - 12 Officers Quarters

- **7** General and General's staff
- **8** Senior Officers
- **9** Officers, Row A
- **10** Officers, Row B
- **11** Officers, Row C (C–5, left center)
- **12** Officers, Row D

13 Generals Lounge / Generals Mess

14 Post Exchange (PX)

15 Corps of Engineers office

16 Vietnamese Gift Shop

17 Storage Shed

18 Officers Club

19 Volleyball Court with Basketball Hoops at each end

20 Parking Area

21 Eakin Compound Headquarters (HQ)

22 Officers Mess

23 Kitchen

24 Fenced-in Receiving Area and Kitchen Barbecue Grills

25 Enlisted Mess

26 Covered Breezeway / Arcade

27 Enlisted Mens Club

28 Chapel (also used as a theater)

29 Multi-use Office Building

- **a** Athletic Director's Office
- **b** Can Tho Mess Association Office
- **c** Red Cross Office
- **d** Post Office
- **e** Barber Shop
- **f** Snack Bar
- **g** Temporary Two-Man Quarters / Storage Room

30 Laundry

31 Storage Shed / Carpentry Shop

32 Fenced-in Storage Area

33 Basketball Half Court

34 Handball Court

35 Compound Supply Facility

36 Transient Billeting

37 Compound Security Barracks

38 - 45 Enlisted Billeting

- **45** Senior NCO Row

46 Senior NCO Club with fenced yard

47 Old Soccer Field, converted to a Heliport after the 1968 Tet Offensive

THE OFFICER'S CODE: DUTY, HONOR, COUNTRY

"Honor is the hallmark of officerlike conduct. It is the outgrowth of character. It means a person who has the knowledge to determine right from wrong, and the courage to adhere unswervingly to the right."

"The essential attribute of the Army and its members is integrity. It is the personal honor of the individual; it is the selfless devotion to duty which produces performance integrity in the discharge of individual missions . . ."

The OFFICER'S GUIDE
1965-1966 edition

MESS

Mess, noun:

2. b. A confused, troubling, or embarrassing condition; a muddle.
4. a. A group of people, usu. soldiers or sailors, who regularly eat meals together.
   b. Food or a meal served to such a group.
   c. A mess hall.

Mess, verb:

—tr

2. To botch; bungle.

—intr

1. To cause or make a mess.
3. To intrude; interfere.
4. To take a meal in a military mess.

THE AMERICAN HERITAGE COLLEGE DICTIONARY
THIRD EDITION

ONE

MY ASSIGNMENT REQUIRES A TOP SECRET CLEARANCE

ANOTHER dreary military tradition is taking place.

Date:	Friday, March 10, 1967
Time:	0900 hours
Place:	An auditorium, Fort Lee, Virginia
Assembled:	One hundred and fifty each almost brand-new Second Lieutenants, Quartermaster Corps, who have just completed advanced training in various specialties.
Purpose:	To inform this outstanding group of their individual duty assignments. They are also to learn about the U.S. Army assigning process. This is a Class 1 uniform occasion.

Including yours truly, sixty-seven of these swashbucklers had had their gold bars pinned on January 19, 1967. This select group had trained together through a grueling six months of Officer Candidate School (OCS). When our class had first assembled on a sunny summer day in August 1966, we had numbered a full one hundred and twenty-five eager young hopefuls, and I mean young: incredibly, at age twenty-two, I had been one of the oldest. Now we were the survivors, and we knew and respected one another. The others in attendance most likely received their commissions from the college program known as ROTC (Reserve Officers' Training Corps). We OCS graduates felt that we were vastly superior military specimens by comparison, but at that point we all shared the same anxiety regarding our first duty assignments. This anxiety was to be abated or abetted.

Colonel Danforth has come down from Washington. Upon his finely squared shoulders rests the duty of assigning worldwide all second and first lieutenants, Quartermaster Corps (QMC), and he knows his subject. Following some refreshing opening remarks that make it clear to us how difficult and how very important are both the Colonel and his job, the floor is opened for specific questions from the assembled. Specific questions he gets, and without using his notes (the Colonel knows his subject), specific answers he gives:

Second Lieutenant Dumas—Kaiserslautern Cold Storage
Second Lieutenant Hirsh—Korea
Second Lieutenant Finley—Hong Kong
Second Lieutenant White—Kansas
Et cetera, ad nauseam.

Some of the fellows even bother to try to argue with Danforth, letting him know that he has made grave errors. But alas, when he is able to give a hapless second lieutenant what is evidently received as bad news, the Colonel actually seems to light up, to twinkle—this is his revenge, his joy. But his fun is running out with each tick of the clock, and the big kicker, the ultimate joy of this trip, is not happening. *Why won't Second Lieutenant Dinan get up off his behind like everyone else and beg me? Me, Mr. Smart Colonel, knowing all and having all of your dear young lives in my control. I, who can make you or break you. Ask for Christ's sake, time is almost out. If you don't ask? Shit! Ah well, on to plan B.*

Audibly: "Is there a Lieutenant Dinan in this group?" *Of course there is, I have committed to memory all the names of all one hundred and fifty assembled. They are all here. I counted.*

Coming to attention: "Lieutenant Dinan, Sir!"

"Lieutenant, why haven't you asked where you're going?" *I hope my annoyance isn't showing too much. It is!* "Is it possible that you don't care about your assignment?" *I think my temper really is showing.* "Well?"

"Sir, I certainly do care very much about my assignment but I knew I would be instructed in due time and therefore I did not venture the question."

Ah, at last . . . Joy truly begins to flow through these veins. "Oh, I see. Gentlemen! Were you aware that our Lieutenant did not care to venture the question?" *Do you believe it? Venture the question! Oh where the hell is my composure going?*

"Lieutenant Dinan, it is my duty—*and pleasure, you little shit*—to instruct you about your assignment."

Resquaring his shoulders and making some show of consulting his papers for the very first time before this assemblage, Colonel Danforth is visibly straining to keep the corners of his lips from curling up. He fails. His eyes are on fire, dancing. This is one happy fella. The tension is almost perfect; you could hear a pin drop.

Now, clearing his throat: "Let me see, ah yes, classified, classified." *It's getting better all the time.* "Oh yes, yes, here we are. Lieutenant Dinan, you will be replacing a senior captain in an advisory team in Vietnam. Please don't ask me any questions about your assignment because the answers are classified. Just one thing: you will need to have your clearance upgraded to Top Secret. Lieutenant Dinan, is there any further question or comment you might wish to, ah, venture?" *What total joy.*

"No, Sir."

"You may be seated."

"Thank you, Sir."

TWO

GETTING THERE IS HALF THE FUN

SAN Francisco can be particularly beautiful in April, and Travis Air Force Base isn't such a bad place. Having a week to kill there before flying to Vietnam can inspire any young officer's imagination. And seeing the plight of the enlisted men who were confined to post, billeted in a gym with army cots stacked three high, and forced to do stupid menial labor to keep busy, almost made the many hardships and indignities of Officer Candidate School seem sweet.

Along with two other second lieutenants-in-waiting, I rented a car, and together we did most of the must-do sightseeing things: the wharf, hanging ten on the trolley, topless bars—hundreds of them. One even had a nude on a swing. No big deal, perhaps, but this was 1967 and the West Coast was very different from the East Coast.

A few hours before boarding, having been cleared, checked, and processed, I finally progressed to the point where I had nothing I had to do. I found my way to the officers bar closest to my boarding area with the intent of contemplating the possibilities of my upcoming adventure in solitude. This was not to be. Evidently, I had the I-am-going-to-Vietnam look imprinted clearly on my face, and this watering hole so close to the area of departure was, not ironically, equidistant to the area of return. The lucky ones, boy, were they having a good time! What camaraderie, what great, genuine, and good-natured happiness abounded, and they were not about to have my imprinted face spoil their fun. They drew me in, and they drew me out, and by the time I had to go, I left filled with joy. I was loaded onto the plane, or was it rather that I got onto the plane loaded?

My joy notwithstanding, 198 men in uniform packaged 3 to the left and 3 to the right over the interminable length of thirty-three rows, bound together for an upcoming twenty-two hours of discomfort on their way to where they didn't wish to go, cannot be a pretty sight. Mercifully I slept; actually I passed out for all of the eleven hours to our first stop: Guam.

What can be said about Guam? Guam is Guam and is so named because, after all, it is Guam. Yet everyone was thankful for the opportunity to get out of the aircraft, move around, wash up, stretch, and so on, all the while trying to prepare mentally for eleven more hours of seemingly endless discomfort. As for myself, once reseated I decided to try counting blessings instead of sheep. I figured the likelihood of sleep was, in any case, highly remote.

You just never know when good fortune is likely to smile upon you. It was Friday night, February 17, 1967. I had graduated from OCS a month earlier, and thus I was happily enjoying the Caribbean Bar, aka the Car Bar, at the FLOOM—that's the Fort Lee Officers Open Mess, aka the officers club. It was about 2100 hours (9:00 p.m. civilian time), and we graduates had been celebrating the fact of our young lieutenantness, which is a rather usual thing for new officers to do. A couple of the fellows had dates, I didn't, but they were girlfriends of long standing and therefore not to be considered in any way available. Then in came Second Lieutenant Rob Dumas, who had been my bunkmate throughout much of our six months of OCS, and with him was a stunning redhead. I knew enough about Rob to know that this was open season, so I turned the old charm up a notch and went hunting. It turned out that Rob had really wanted to introduce Anne Kinsey to me because she was majoring in History of Art, and I had kept a postcard gallery of great art works taped to the inside of my personal locker door. This was considered quite weird, as one should really have naughty posters, Playboy centerfolds or some such. Anyway, thinking that we both had, at the very least, this little strangeness in common, he figured we should get to know one another.

As for the hunting, I didn't score a kill, but I guess I didn't shoot myself in the foot either, and so I could count myself lucky to have a girl back in America to write to and think about. It was a blessing. I had seen Anne just before heading west. I had borrowed Lieutenant Wisemendel's smashing new car and driven to Sweet Briar College from Fort Lee, attired in my beautiful, new, and very impressive, uniform. When I knocked on the door of House One, a very attractive young student opened the door with a cheery bright smile and inquired with all her southern charm intact: "Are y'all here to read the water meter?" So much for the impressive uniform! Anyway, we had a great day on that beautiful campus. I knew I was lucky to have things like that to reminisce about.

"This is the Captain speaking. May I have your attention please?" We really couldn't say no but, in fact, after ten hours in the air, I was ready to give my attention to anyone with something even remotely interesting to say. "I'm sorry to inform you" (*hold it, wait, you have my attention but don't start getting sorry to inform me now that I'm so close to terra firma*) "we have received reports of extensive shelling at Saigon's Tan Son Nhut International and we've been diverted to Cam Ranh Bay. We hope this doesn't inconvenience you too much, and once again I want take this opportunity to thank you on behalf of the entire crew for flying Braniff."

A loon! We've been flying with a loon! To this boy from the Bronx, extensive shelling translates into they are dropping bombs all over the bloody place, killing and maiming people—and this is the place where I'm supposed to be going. And one of those people is supposed to be me, and this loon hopes I'm not too inconvenienced and takes the opportunity to thank me for my being tortured in this crate for twenty-two hours? No! No! This is all too much to be true. Let me out of here! I promise if I ever get to walk off this plane, I will kiss the ground.

And kiss the ground I did. Right down on all fours I went, right there in front of God and everyone. I got the feeling that this was not a usual occurrence but I could tell that those who had been on the plane with me understood. As for the other spectators, well anyway, they had to get on with the business of being official, and dealing with this sort of behavior was not in their scripts.

For all the French influence in Vietnam, and at that time there was still plenty, high noon at Cam Ranh Bay International in April is not your basic April in Paris experience. Temperature? 120°F easy. Shade trees? Not easy. Our official welcoming party: a sergeant with a clipboard in his hand and JACKSON on his shirt. "Welcome to Vietnam gentlemen. Would everyone assigned with USARV (United States Army Republic of Vietnam) report to the mess tent on your right? You will begin processing after lunch."

One hundred and ninety-six hot and hungry raised one hell of a cloud. When the dust settled, there we were, Captain Youngblood and Second Lieutenant Dinan, both assigned with MACV (Military Assistance Command Vietnam).

The Captain, taking charge: "Sergeant Jackson, what are our instructions?"

"Oh, you're with MACV. You've got to get to Saigon."

"How do we go about doing that?"

"Gee, I don't know."

"What about a little lunch before getting onto that business, then?"

"No. I'm sorry. We're USARV; we're not authorized to feed MACV."

"I see," replied Captain Youngblood. He saw, I didn't. He continued, "Well then, is there a snack bar near here?"

"Yes, but you don't have authorized currency, you know, MPC, military payment certificates."

"Sergeant, where do we get this MPC?"

"You get that when you process with your unit."

"Sergeant Jackson, will you please escort us to your unit's officer-in-charge?"

"Well, Sir, you see . . ."

"No, Sergeant, *you* see. I am giving you a direct order. You will escort us to your OIC now!"

By now, Youngblood's blood was really boiling and getting hotter all the time. By the time we were turned over to the OIC, his veins were bulging such that I thought he was going to have a stroke. Thankfully, this OIC was a lieutenant and therefore Youngblood was able to avail himself of that time-honored military tradition of pulling rank, thus relieving one's tension and stress by giving it to a subordinate.

Truly the man was upset. "Goddamnit, Lieutenant Adams! You stand at attention when I address you. What the hell kind of a shithouse are you running here? What's all this bullshit about USARV and MACV? Are we Americans or what? I want that Sergeant brought up on charges. Who do you report to? How do I get fed? How do we get to Saigon from here? Walk?"

And he went on. Lieutenant Adams was turning white, but Captain Youngblood's veins had returned to normal along with the pitch of his voice and his general composure. Then, in a conversational tone, "Lieutenant Adams, I know this is not your fault. Forget most of what's been said and see what you can do to be helpful to us."

"Yes, Sir." The boy definitely did not want to get this volcano reactivated. "Please, if you will come with me, I'll see to it that you get something in your stomachs. I know it sounds crazy, but for the most part what Sergeant Jackson told you is true. United States Army Republic of Vietnam has almost nothing to do with Military Assistance Command Vietnam. You're on COLA, and therefore we can't draw rations on you, which translates into we can't feed you except in an emergency."

So I ate my first meal in Vietnam as a charity case. Somehow, I felt something was weird.

After lunch, we were taken to the air terminal that handled in-country flights. We showed our orders to someone in charge who told us he would get us on a flight to Saigon as soon as possible, but since we weren't manifested he would have to squeeze us in. He warned us to be prepared for a long wait. So, at length, I had my first flight in Vietnam as a charity case. Somehow, I felt something was weird.

When we arrived at Tan Son Nhut, it was dark. You could hear the sounds of gunfire all over the place: automatic weapons, single shots, mortars, bombs. I couldn't tell if they were near or far away, but no one seemed to pay much attention, so I figured it was normal. Captain Youngblood and I were in Class 1 uniforms, the same ones we were in when we boarded the Braniff flight at Travis Air Force Base, half a world away in California. Everyone else was sporting some sort of combat dress; most were carrying one sort of weapon or another. We were carrying footlockers and duffel bags, so keeping up with the crowd was anything but easy. By the time we got into the terminal, everyone had dispersed. We had thought there would be someone present to provide information on how we were to proceed to MACV Headquarters. We were wrong. It was late, and the place was closed up. No one was there except for a few Vietnamese, both in and out of uniform, who couldn't speak English and a few enlisted men who knew nothing. There was a telephone. Captain Youngblood seized it like a long-lost friend, and after about two hours of relentless dialing, waiting, and redialing the next recommended number (I can't help you, try this, try that), these old friends were over-reunited, and a body was on the way to pick us up.

After reconnoitering the area, we decided that there were three logical spots, each one out of sight from the other, where a vehicle might be stopped to facilitate picking us up. So we had to keep moving from one spot to another, watching for our ride, which we assumed would be a military vehicle. After what seemed too long a waiting time, we began to notice a light gray pickup truck that had been around for a while, unusual because all other vehicles had come and gone rather quickly. Getting excited all over again for the hundredth time like a couple of strandees on a desert island, we quickly went to inspect this, our greatest of prospects. Nothing, empty, no sign of life, no shred of evidence. As our spirits had gone up, so our spirits now went down. Depressed, we decided to try another telephone call. As we headed toward the phone, we noted a Vietnamese man dressed in civilian clothing, just a shirt and trousers and the obligatory Ho Chi Minh sandals. There were others around dressed similarly. The thing that made this fellow stand out was that he really seemed to be aware of his surroundings and seemed alert. He was actually looking around. We figured, no, it can't be, it would be too out of character for this day. Desperate, we made a closer inspection anyway. He was holding onto a well-smudged piece of paper on which I could make out clearly the word DIAMOND. All else seemed impossible to decipher. "Captain Youngblood," I said, "this is for us."

"How do you figure that when there is nothing that says so?"

"Look! His paper says DIAMOND." The Vietnamese man knew he was part of our conversation, but from the look on his face, it was clear he didn't understand one word being spoken.

"I can read, Lieutenant Dinan. It says DIAMOND. So what?"

"No one, I mean *no one,* ever gets my name straight on the first try. I've been called everything from Dino to Dinian and I get ten Diamonds to every one Dinan. So please, let's give it a try."

As we approached what we hoped was our driver, he didn't become real friendly. I pointed to his note and then to my name tag, giving the universal body language YES with my head. Meanwhile, he was not to be fooled. After all, even if he couldn't speak our language, Vietnam has the distinction of being the only Asian country to share our alphabet, and even the uneducated can plainly see that DIAMOND is not DINAN. So he was simultaneously giving me the universal body language for NO and identifying the proper party he sought by pointing to each letter, one at a time, with his forefinger. Clearly, he was the teacher, and I, the student. Seeing an impasse but remaining hopeful, Captain Youngblood took a piece of paper. On it he wrote MACV and showed this to my teacher. Something clicked, and right away it was all smiles and "*Choi oi*" (translation: "What the hell"), so we followed him to the light gray pickup truck. I got in next to the driver, giving the window seat to the Captain. Thanks were not required; it was his due.

I'm not sure if our driver was good. I am sure that he was fast. Like a bolt, we were shot into the midst of teeming Saigon traffic. It was almost midnight, and the city was very alive. The streets and boulevards were filled with every imaginable form of locomotion: mopeds, cyclos, bicycles, pedicabs, little taxis, cars, pedestrians, the works. It was a jungle, survival of the fittest, and we were the biggest, therefore the fittest. With one hand on the horn, the other on the wheel, head out the window screaming what I suspect were uncomplimentary epithets, and traffic lights ignored as if nonexistent, our driver careened, braked, accelerated, swerved, and bolted through the crowds. White-water rafting is child's play by comparison. There was no stopping him, and I couldn't get the thought out of my head: What will they do to us when he runs over a half-dozen or so of his countrymen? He'll disappear, probably a VC (Viet Cong) anyway, and we'll be torn limb from limb by an angry mob. It was awful—the worst, the most frightening ride of my life. Incidentally, I was later to discover that the traffic lights had just been installed two weeks before this midnight ride, and it seemed a matter of national pride to pretend that they were not there.

As quickly as we had entered into the stream, we were out of it. A sudden, screeching left turn hurled us from the light into darkness, and mid-block we slammed to a stop. Happily, we got down from the cab. Still shaking, we gathered our belongings from the flat bed of the truck. When our kamikaze figured we had it all, the accelerator hit the floor, and with the skill of an American high school senior, our driver laid down a patch of rubber that was second to none. We were alone in the dark, mid-block, somewhere in Saigon. The only sounds heard were the not-so-distant sounds of war.

As our eyes became adjusted to the darkness, we found that we were standing in front of a wrought iron fence. A few feet from where we were standing was an opening in the fence, and just inside was a little guardhouse. Standing next to the guardhouse was a uniformed soldier complete with rifle. He was small and watching us carefully. When we approached, we noted immediately that he was Vietnamese and a friendly. Unfortunately he did not speak English, but MACV was again understood, and with that password, we gained admittance. Across the small courtyard was a yellow brick building some five or six stories high. There were no lights on, but the door was unlocked, so leaving our baggage behind us, we went in to explore. We went up and down and all around and found no one. The place was empty.

Back outside, we sat down, lit up, and started to speak at the same moment. "Excuse me, Sir. You first," I said.

"That's not necessary, Dinan; you have your say. I think I might have said my say for today much earlier."

"OK. This is not possible. It simply is not possible. I think our little ally in the guard booth knows where we should be going, and even if he doesn't speak our language, he could help us. I bet he can't wait to tell his crowd about the two winners that showed up on his doorstep via the pickup truck express. I can't believe that we would be expected to hang out in this empty building or courtyard until daylight."

"You know what, Lieutenant?"

"What?"

"Neither can I. What say we ditch these cigarettes and recon the place?"

"Fine by me."

We headed to our left, clockwise around the building. Turning the first corner: nothing, blackness. We kept going. Turning the next corner, we could see a light up ahead and over to the right. We were drawn to it like flies. Lo and behold, an office. And in the office we could see an officer, an enlisted man, and a television. Ah! Civilization. It was the officer of the day (OD), the one who stays on duty throughout the night, and his charge of quarters (CQ), the enlisted man who gets to keep the OD company. They weren't real happy about having their TV program interrupted, but the OD was a lieutenant and Youngblood was, thankfully, a captain, and on his coattails of power, I too was quickly assigned a bunk. But not before receiving the welcoming instructions: wake-up was normally 0600 hours, but since, by arriving late, we had missed the opening of our in-country briefing cycle, we could sleep as late as we wished. The next series would begin on Wednesday. It was now 0130 hours Tuesday. There was a money exchange on the compound where we must exchange all our U.S. currency for military payment certificates during regular business hours. With the MPC, only, we could buy merchandise at the snack bar or the PX. We couldn't have any Vietnamese piasters until we received the proper identification card, which you couldn't get until after the second day of in-country briefing, which didn't matter because you can't leave the compound until then anyway. GREAT! The only places I wanted to go to were the shower and the bed. It had been a long time since I had seen either, and I was determined to make up for the loss. Not knowing what the immediate future held in store for me and guessing that it might be rather grim, I determined to take full advantage of any luxuries available. So I slept in until 10:30 (1030 hours) and then took a very long, serious hot shower.

By the time I dressed and changed my money into MPC, it was noon. I wanted to avoid having to answer questions about where I was going until after I found out the answer, so instead of going to the mess hall, which was crowded for lunch, I went to the snack bar, which was empty. I hadn't eaten in about twenty-four hours, but somehow my appetite wasn't inspired, so I just got myself a cup of coffee and sat down, thinking to take inventory of the last two joy-filled days. Just then the door opened and in walked a civilian wearing a madras plaid sports shirt and pale blue chino-type slacks. He had a great tan, sun-bleached hair, and the beginning of a beer drinker's paunch. I figured him to be about thirty-five and a hustler. It took a minute for his eyes to adjust from the brightness outside, but you could sense by the way he waited that he was looking for something or someone in particular. I watched as he started to survey the room from left to right, and when his eyes landed on me, it was obvious that I was that someone.

But this didn't make sense. He couldn't have read my name tag from that distance. I prepared for a case of mistaken identity as he barreled down upon me. Reaching out his hand and looking me straight in the eye, not at my name tag, he said, "Welcome to Saigon, Lieutenant Dinan."

"Oh, thank you. I do believe that's the first personalized welcome I've received since my arrival."

"I thought you were supposed to get here yesterday."

THREE

THE INDOCTRINATION

AT that particular moment, I was thinking, who is this guy? He knows who I am, he knows when I should have gotten here, but he doesn't tell me who he is so I figure, inasmuch as he's a civilian anyway, I'll just play along with this coolly and see where it goes. "Actually, I was supposed to get in yesterday, but we were diverted to Cam Ranh Bay. The captain of the airplane said, 'We hope this doesn't inconvenience you too much.' If only he had known!"

"Yeah, yeah. It can be tough getting around the country, especially with a lot of baggage."

How does he know about my baggage? I'm not going to give him much to go on. "Quite true," I said, "but ah, look at the beautiful day we have today."

He contemplated for a moment and replied, "Say listen, are you doing anything?"

"No, not really."

"Why don't you and I go and have lunch away from here? We have to talk, and I know a great little French restaurant not far from here."

I don't even know this guy, and he wants to take me to a great little French restaurant? I don't think he's gay, maybe CIA. *We have to talk*—could this have something to do with my assignment requiring a Top Secret clearance? I don't know, but I am getting a little uncomfortable. "That sounds great, but

I'm confined to post until 1700 hours Thursday afternoon." I wasn't going to tell him why.

"Oh yeah, the briefing business. Well then, we'll just have to do something to celebrate your freedom then. Say dinner, at the EM club. I can't go to your club, but you can come to mine."

Wait a minute? Did he say EM Club, as in an enlisted mens club? "What club is that?" I asked.

"We could go to any of them. There are five big ones here. Actually the EM clubs are better than the O clubs. There are more of us, and we spend more. I hope you don't mind my saying so, but officers are cheap."

I do mind, but I can deal with that later. So my civilian/CIA big shot turns out to be an enlisted man; it's time to change tactics: "We'll see. Tell me, how did you know who I am?"

"As soon as I saw the gold bars I knew it had to be you. There just aren't that many second lieutenants that come through this place."

"I see. What is it that you do here?"

"Well, that's part of what we have to talk about."

"What's the other part?"

"I'm supposed to tell you about your assignment."

"Who do you work for?"

"Captain Lane."

"Who's Captain Lane? Never mind, do you happen to have any identification?"

"Oh yes, sure I do."

Something about this guy was really starting to annoy me. You would think he would offer to show it to me when I asked. It's like he wanted to be thought of as my equal, maybe superior.

"Let me see it, please." There was an edge to my voice.

"Gee, you don't have to go getting so official on me."

"Yes, I do. Now."

I was presented with an EM ID complete with picture, thumbprint, rank, and name: Sergeant E-6 Michael Blevens. My assignment, my assignment.

He said he was supposed to tell me about my assignment. What a country, the unbelievable keeps happening. This hustler civilian, could-be CIA big shot, active Sergeant E-6 in inactive sportswear who wants me to be one of the guys; this guy has been entrusted to divulge to me my Top-Secret-clearance-required assignment. We're not going to win any wars this way, that's for sure.

"Thank you, Sergeant Blevens. Is there any reason why we can't talk right here?"

"No, but . . ."

"No buts, then, please. Believe me, what matters most to me is finding out what my next year is likely to be like—so give."

"OK, it's like this. You are going to replace Captain Lane as the Custodian of the Can Tho Mess Association, and believe me, he's anxious to get you down there so that you can sign for all the inventory and stuff and he can go home. You have some mighty big boots to fill if you ask me. No one here can figure how the geniuses in Washington decided to replace a senior captain with a second lieutenant anyway."

"Please Sergeant, never mind the big boots and the geniuses. Tell me, where is Can Tho?"

"Can Tho is right in the center of the Mekong Delta. Actually, it is the capital city of the Delta, and that's where the Vietnamese Military Headquarters is located, and we, MACV, are there to advise them."

"So, I'm going to be advising."

"Oh no! You're going to be the Custodian of the Mess Association for Eakin Compound; that's the billeting compound for MACV IV Headquarters."

"Is it dangerous there? I mean . . ."

"I know what you mean, Lieutenant Dinan. The answer is, it's not the worst. The Viet Cong rocket and mortar the place every so often, which makes for a lot of purple hearts, but I haven't heard of anyone getting it posthumously lately."

"That's cheerful news. Now, what is this mess association?"

"Eakin Compound houses the Commanding General of MACV IV Corps, Brigadier General Desobry, and the General's staff, officially designated as Advisory Team #96, altogether about three hundred officers and enlisted men, all of whom are on COLA, so . . ."

"Wait, the guy at Cam Ranh Bay told me I was on COLA, but I never found out what that means."

"It's cost of living allowance. It's used when Uncle Sam figures it's cheaper and easier to give the boys out in the field money instead of rations. He says, 'Here's some money fellas. You figure out how to get yourself fed.' So everyone pays into the mess association, and we buy and cook the food."

"All right. Now, you said I am going to be the custodian of this association. Growing up, the custodian was always the guy who cleaned the school buildings. Do I take it . . . ?"

"Wait. No. You are the custodian of the funds. Didn't they tell you that officers don't work?"

He waited for me to show some sense of humor. I didn't. "Speaking of work, Sergeant, what is your job and why are you not in uniform?"

"Getting official again, huh?" When he figured out that I was not going to respond to that question he continued, "OK, Sir. I'm the buyer. It takes a lot of food to feed three hundred people three times a day, seven days a week."

"Fifty-two weeks a year," I added.

"That's right, Sir. Also, I do the buying for the clubs there run by Chief Warrant Officer Arnold, great guy. First I secure the merchandise, and then I arrange to have it flown or shipped down. Actually, we only fly it down now because we don't have the boat anymore. We have some storage space assigned at Tan Son Nhut, and when I get a plane load assembled, I arrange for air shipment to Can Tho. There's a lot of politicking involved, but I know the people, and I get the job done."

Obviously this guy is trying to impress me, and he's not doing a bad job of it at that. "I see. That sounds like a big job, but what about the clothes?"

"Well, I've got to live off the economy while I'm in Saigon, and it's safer in civvies than in uniform."

"Doesn't that get expensive?"

"Between the COLA and the per diem, it works out."

"Who pays the per diem?"

"The mess association."

"And how long have you been doing this?"

"Ten months down, two to go."

"When was the last time you were in Can Tho?"

"That would be the last time we brought the merchandise down by boat, about six months ago."

Six months? I wonder what kind of a scam this must be. This is quite a guy, my Sergeant: a free agent on the U.S. Army payroll. I wonder how many more I have like this one. "You certainly must do a lot of buying."

"Oh yes, I sure do. Listen, why don't I pick you up here at 1700 hours on Thursday? You'll really want to get away from here by then, and a good steak will do you good."

Not being able to think of any good reason to say no, I replied, "That sounds fine to me. 1700 hours it shall be. Now I suppose you must get on to buying, and I have some letters to write, so, well, thank you for coming over here to see me. It's been very nice to meet you."

Standing up, I extended my hand to him, and taking his, I walked him to the door.

"Oh, ah, yes, yes. Thank you. It, ah, was great to meet you too. 1700 hours then." Sergeant Blevens was a little taken aback. He hadn't expected such an abrupt ending to our meeting and I suspected that he wanted to score a few more points on the importance of his staying in Saigon, but I wanted to make sure I established the right relationship with my enlisted personnel. After all, I was a brand-new second lieutenant on his first assignment, and I had heard the usual horror stories of how these pros can eat you alive once you've gotten off to the wrong start. And in my naiveté, this guy looked like a real pro.

Now my appetite became truly inspired, so I treated myself to a cheeseburger, French fries, and a coke. Captain Youngblood came in, and he didn't look good. He walked over and sat down. Close up, he looked worse. There was something really odd about the way he looked. I couldn't figure it; even his voice sounded unnatural when he asked, "What have you found out?"

"Did you notice a civilian in a madras shirt and light blue pants?"

"I think so."

"That was my buying sergeant—claimed he would be unsafe in uniform. Anyway, he told me that I'm going to Can Tho, down in the middle of the Mekong Delta. It seems that I'll be running some kind of food operation at the General's compound, a place called Eakin Compound. This guy's been connected with the operation for ten months, and he says he hasn't heard of anyone getting killed there yet, so I guess I'm making out pretty well. Any news on your assignment?"

"I'll be heading up the other way to take over a five-man team out in the field in I Corps sector. The man I've been sent over here to replace was killed two weeks ago. I don't have to guess how well I've made out." And he had that look, a look I couldn't identify because I hadn't seen it before: *fear*—cold raw fear, the certainty of death. I didn't know what the hell to say. I felt like hiding my face. I felt ashamed of my good fortune, but I felt glad that it was he instead of me. What could I say? I answered, "That really sucks."

"Agreed."

Youngblood got up and walked away. He never looked back, and we never passed another word. The odds are that when he returned to the United States, if indeed he ever did, he probably traveled in a plastic bag and that sucks—agreed? That really sucks.

The in-country briefing was interesting but unmemorable, the expected at every turn. As advisers, we were to show particular interest in Vietnamese customs and mores. On the other hand, we were advised not to show any interest in Vietnamese ladies. If invited into someone's home for a meal, you must say yes, and you must go. To do otherwise would cause tremendous loss of face for the Vietnamese. But by all means, never ask what you're being fed. After all, it could be the family dog or cat that has been sliced up in your honor. Therefore, asking would be very rude, so eat, lose your lunch, but do not cause your host to lose his face. Be careful of the water. Take your malaria pill weekly. Use salt tablets. Keep your feet dry, and above all don't overpay for goods or services. The Vietnamese government has let us know, in no uncertain terms, that the American soldiers are ruining their economy by overpaying for everything—something the French never did. So now the Vietnamese themselves can't get a taxi or a table in a restaurant or a room in a hotel or afford the goods on the black market. It struck me as strange that a government would complain that its citizens could not afford to buy on an illegal market, but for my part I wasn't interested in overpaying for anything anyway.

By the time 1700 hours Thursday rolled around, I was equipped with my in-country ID, my MPC, and my Vietnamese piasters. The "money" all looked like something out of a Monopoly set to me. So to keep the currencies separate, I put all the MPC into my right pocket and the piasters in my left. Thus equipped, ready for that steak, and ready to break out of this little jail of a compound and experience some of the greater sights of Saigon, I went to the snack bar to find my Sergeant and guide, Sergeant E-6 Michael Blevens. Surprises, always surprises! What do I find? Blevens had gone military on me. There he was in full tropical fatigues complete with all allowable notices of military accomplishments. "I like the uniform, Sergeant Blevens, but why? Won't you be unsafe like this?"

"Well, you see, we—you and I—we'll be going to a senior NCO club, and it's better this way." I really didn't see, but on the other hand I really didn't care either. I figured the pro had to have something up his sleeve.

"For me, it's fine but why do I get the feeling you have something up your sleeve?"

"Did you say sleeve?"

"I did."

"Why would you ask if I have something up my sleeve, Sir?"

"That's just an expression. I'm trying to figure out what went into your decision-making process when you were deciding which *dress* to wear for our date tonight." I hated myself immediately for saying that to a subordinate.

This guy had a way of getting to me. I'll have to be more careful. "Forget I said that, Sergeant."

"Sure, it's okay Lieutenant Dinan, but remember, in this man's army we don't put it up our sleeves, we put it on our sleeves."

I felt a light going on, and now I really looked at his sleeves for the first time, and there it was: what was up his sleeve was also on his sleeve. You've got to count those stripes—three up and two down add up to Sergeant E-7. Why three plus two equals seven is a military secret. So secret that even an officer with a Top Secret clearance is kept from that knowledge. The ID said E-6, but the penalty for impersonating a superior rank is so severe that not even this soldier would risk it for the sake of one grade, so I concluded that he probably hadn't gotten his updated ID, most likely because he hadn't been to Can Tho in six months. But why? Why all the subtlety? I figured, ah, why not? I'll be subtle too and pretend not to notice.

"OK, you look great. Now let's go get those steaks."

Turning his arm toward me to be sure I got a full frontal of his stripes while watching my eyes (you can bet I kept them tight on his), he answered, "Yeah, let's go. What say we take a cyclo? It's not as fast as a taxi, but you can see better."

"I'm in your hands."

So off we went. I felt a little lightheaded as we left the compound. You know, it was the first time I had the feeling of an army officer abroad on the loose in a foreign country, in a foreign city, *their* city. The sun was brightly shining, and everyone seemed to be going about his business—business as usual. As we walked along, no one seemed to pay us any attention at all, and I really felt very safe indeed. We—that is, Sergeant Blevens—hailed one of the three-wheeled motorcycles. Two wheels up front like a delivery bike, but instead of a big basket for goods, there was a comfortable seat for two. With no driver in front to block your view, it was the perfect vehicle for sightseeing. There were more sights to see than I could absorb. The contrasts of young and old, rich and poor, ancient and new, traditional and contemporary were everywhere and in everything: cars, bikes, trucks, buildings, houses, signs, roads, shops, and most of all, the people themselves. New York had nothing on Saigon as a city of contrasts. Thus distracted, I had no idea of how we achieved our destination, but there we were in front of what had obviously been a luxury hotel, now guarded out front by sandbagged machine gun emplacements rather than uniformed doormen. The building had become a residence for senior U.S. Army enlisted personnel, E-6 sergeants and up. The club occupied the entire top floor, and when you stepped off the elevator, you stepped right into it—and it was plush. Off to the left were the pinball machines, dozens of them. Beyond them, slot machines, hundreds of them, and all in use—a training ground for Las Vegas. To the right was a pool room with half a dozen tables, also all in use and well attended on the sidelines. Straight ahead was the bar. We're talking big, and we're talking truly plush. This baby wove its way into the distance like some great snake. Its rail of deeply padded, soft red leather was backed up by a richly lacquered black bar. The back of the bar was divided into work stations every few feet. Each station was identically stocked with all the brands of everything I had ever heard of, as well as its own bartender: a young American soldier in white shirt, white jacket, black pants, black shoes, and black bow tie. There were mirrors and fancy lights and recessed lights and hundreds of speakers delivering an even volume of music. And in front of the bar were big comfortable bar stools resembling club chairs jacked up off the ground. The carpeting was thick and soft, the dining tables were exquisitely set with starched white linens and gleaming crystal, and beyond the tables, floor-to-ceiling glass formed the walls on three sides. This must have been the original "Windows on the World." From the ceiling hung magnificent chandeliers, and in the center was a stage, almost in the round. I was impressed. I was floored. Someone had spent a fortune putting this act together. The overhead had to be unbelievable. Sergeant Blevens knew he had scored on me, and he was enjoying it completely.

"I hope you like our enlisted club here, Lieutenant Dinan. You're always welcome as long as you're accompanied by a senior NCO."

"War is hell," I answered. "How do you all deal with this sort of deprivation, Sergeant?"

"It took some getting used to but I've managed. Shall we ease over to the bar for a drink?"

"Indeed."

"That's my favorite bartender, the eighth one down. Come with me." I did, and as we sat, this clean-cut looking youngster greeted my host with warmth and obvious respect. It was also evident that he was not impressed with his guest, *me* (being a second lieutenant can be a trial). I ordered a Ballantine scotch on the rocks; Blevens, a CC and ginger ale, a real EM drink, I was to find out. The drinks were big and beautifully served. Our bartender rang up the check and placed it politely in front of Sergeant Blevens, upside down.

I looked over at the register and saw $0.50. "Is that what my drink cost?" I asked incredulously. "Fifty cents?"

"Hell no! Are you kidding?"

I wasn't kidding, I was hopeful. "I'm new in town; I don't know these things."

"Well, you should know that if anyone tries to get you to pay fifty cents for a drink, you're being taken for a ride."

What's he saying? Fifty cents is too high? "Okay, then. How much is it?"

"This, this is twenty-five cents and that's the max."

This information had me dumbfounded. "Tell me how, please, does twenty-five cents for a drink like I have here begin to pay for a setup like this?"

"It doesn't."

"All right, then. What does?"

"The pinballs and the slots."

"The pinballs and the slots?"

"That is correct. Those machines pay for everything here even though they have an 85 percent payback."

"That's incredible."

"Look around you, Lieutenant. What do you see?"

"I see a bunch of senior NCOs who seem very well mannered and very professional."

"That's correct. These are the pros. All these men have got a lot of time in, and they have very important jobs over here. Most are from the old school and single under the prevailing wisdom: 'If Uncle Sam wanted me to have a wife, he would have issued me one.' Just about all of these guys have been busted more than once for being drunk and disorderly, and now they are getting too close to retirement to risk another demotion. So this is what they do with their money. They invest in the slots—not a bad investment when you figure the returns, the food, the service, the surroundings, and the live show six nights a week. It's like you said, Sir, war is hell, but we try to make the best of it."

What a pro. No wonder he wants to stay here instead of commuting to Can Tho. Living off the economy—right!

"Shall we check into those steaks you mentioned earlier this week?"

"Good idea."

"I'll pay for these drinks."

"No you won't, Sir. Officers' money is no good in here, isn't that right, Private?"

"That's right, Sergeant!" replied the bartender.

Sergeant Blevens led me to a table that would have a good view of both the world outside and the stage. The sun was going down fast, and for a short while the sky was brilliant orange, then black, and then lit with randomly located flashes and the steady sounds of war. Ratta-tat-tat, thump thump, ratta-tat-tat, thump thump, an unending, almost soothing background noise that you don't pay much attention to but are nonetheless comforted by.

The meal was excellent and well-timed so that just as we were being served coffee and brandies, the floorshow was getting under way. A well-meaning group from the Philippines was doing a pretty good imitation of a Western-style night club act, not bad, not bad at all. This was definitely the right way to spend a war.

All at once, the music stopped and the lights went up—and now I could hear the sounds of war again, only someone had turned up the volume, and they no longer gave comfort. An announcement blared through the speakers from I don't know where: "Gentlemen, the firing is approaching this sector. You are all asked to settle your bills and move on to your drill stations. Anyone not billeted in this building is advised to leave as quickly as possible."

My brain was racing, and my heart was beginning to pound. I looked to my pro for help, and he was as cool as can be. "No sweat, Lieutenant. This happens all the time, but we do have to get out of here."

We settled up quickly and headed down in the elevator. In this safe cubicle Sergeant Blevens suggested, "I need to go in the opposite direction from where you're going, so why don't I organize transportation for you and then we can go our separate ways?" And then the zinger, "That is, unless you think you'll be uncomfortable alone."

"Sounds reasonable to me." Of course, I'll be uncomfortable alone. But to admit it?

Out in the street we found another three-wheeler with the seat up front. Sergeant Blevens talked to my driver, who looked like a thug, and the Sergeant was satisfied that this driver knew where to take me. "It's all arranged, Sir. Nguyen here will take you right home."

"Great! Thanks. By the way how much should this trip cost me?"

"Give him sixty, no more."

"Give him what?"

"Sixty, that's piasters, their idea of money."

"Right. I should have figured that out for myself. Good night and thanks for the evening." There were plenty of street lights all around when I first got into the cyclo but all too soon it was quite dark. I don't know how the driver could see where he was going. For most of the trip, I could only make out vague images, and I was up front. The sounds of war seemed to be getting closer, and they had become anything but a comfort to me. Ratta-tat-tat, thump thump. What was that over there? Where the hell am I? I don't remember being anywhere around here before. Ratta-tat-tat, thump thump. Where is this guy, this thug, taking me anyway?

I turned around and asked, "Hey you, where are we going?"

All I got in answer was a smile with lots of gold teeth. Ratta-tat-tat, thump thump. Come on, this is bullshit! Man, have a heart. It's my first time out. Where's he taking me? My heart was really thumping now, and sweat was rolling over me. I considered jumping out and making a run for it, but to where? We went over some railroad tracks and made a left. On my right, buildings, shacks, no lights. They seemed very ominous. On my left, tram tracks. Things were scurrying about—rats, not here, no, what? Does this guy behind me have a gun, maybe a lead pipe? There was a street lamp up ahead. We stopped. I recognized the street. Home! Thank you, thank you! But come on, make a right and take me to the gate. The universal NO. This one was going nowhere, just waiting to be paid. What was it? Oh yes, sixty. I reached into my right pocket, counted out three twenties, and handed them over. Quickly moving to the beat of my heart, I headed down the darkened street, right down the middle, ready to ward off any sudden attack. Mid-block I reached the gate. Thank you, thank you. Ratta-tat-tat, thump thump.

Once inside the gate, the feeling of safety began to replace the strain of fear, and quickly shame started to take over. What a jerk! What the hell were you so afraid of? Some ninety-eight-pound little Asian that you could take out with one punch? I mean, man, you had best get it together or you're going down as the coward of the century. But wait, wait just a minute. I'm no coward. There must have been a good reason for my terror (we're so good at making excuses for ourselves). The facts. Let's look at the facts: It's pitch-black. I'm in a strange city filled with a palpable energy I can't identify. There are bombs going off all over the place. And I sit all alone, unarmed, with my back to a stranger, who looks like a thug, and who could just as easily be an unfriendly (Viet Cong) as a friendly. And I ask why I'm scared? Good grief! What I should ask is, why am I so stupid? What a jerk! This is a Paul Revere special I don't intend to repeat.

I felt much better now. It seemed a cold one was in order, so to the snack bar I went for a Budweiser, the king of beers. I reached into my left pocket and came up with a handful of Vietnamese piasters. Wrong pocket. Oh No!

Reaching with my right hand, I retrieved a seriously depleted bankroll of MPC. Oh! Jerk! Schmuck! How in the . . . damn! I am so pissed, I can't believe it. This streetwise young officer from New York City has just laid out $60.00 for a sixty-cent thrill ride of terror. Shame, shame, shame on me. That does it. I am not leaving this compound again until I have to. Damn!

The next couple of days went by rather uneventfully. By 0700 hours Monday morning, we had completed our in-country briefing cycle. Now we would be issued our gear, and we were told to depart the compound not later than 0900 hours to make room for new arrivals. The gear consisted of four sets of jungle fatigues—green, two pairs of jungle boots—green, six tee shirts—green, six boxer shorts—green, six pairs of socks—green, and a baseball cap—green. I was asked to accept a helmet and helmet liner but declined. The issuer let me know that he thought I was more than a little deranged, but I tried to explain that whenever I had to wear one of those five-pound monsters, my neck would hurt and I'd get a headache, so I just wasn't interested. He answered with something stupid like, "Wait until you get your head blown off; then you'll know what a headache is."

"I'll give you a full report," I replied, and then it was off to weapons. What an arsenal. These guys had everything, even crossbows.

"What's your favorite, Sir? M16, M14, carbine, grease gun, Thompson. You name it, and most likely we've got it."

"You mean you'll just give me any weapon I decide on?"

"That's it, Sir."

"In that case, I'll take the smallest, lightest thing you have."

"Too bad. We just issued out our last .38 Saturday; we'll have to give you a .45."

"That will be just fine, thank you."

I was issued a .45-caliber automatic pistol complete with holster and ammunition. The fact that I had never before held one in my hand didn't seem important to anyone. The important thing was to sign the correct forms with the correct numbers so that accountability for its possession would be properly transferred. Its use? That was a personal problem.

Thus armed, along with my footlocker and two duffel bags, I was now ready to head south to Can Tho. Sergeant Blevens picked me up in a jeep he had managed to borrow. At this juncture, those assigned to MACV were evidently expected to get about on their wits. The ride back to Tan Son Nhut airport was pleasant enough. After all, the sun was shining, and although the temperature was steaming, the breeze was pleasant—well, almost pleasant. The air was ripe with the smells of rotting things. Some areas were worse than others, but the smell was with you always. I hoped that it would be less odorous away from Saigon.

We went to the air cargo location. I just can't bring myself to call it a terminal. It was a shack surrounded by CONEX military shipping containers, some of which were refrigerated, all of which were assigned. Five were assigned to the Can Tho Mess Association, and of these, two were freezer units. This was where my Sergeant assembled the goods before shipping them south, and for whatever reason, he was a very welcome visitor indeed. All the senior NCOs there seemed to show him too much respect for my liking; on the other hand, they seemed to show me too little.

"Yo! Blevens! How's it going? This must be your new second louie?"

"This is him, boys. Lieutenant Dinan, you'll have to get to know these guys and be good to them; they're an important part of your life line." Without really introducing me, he went on, "Captain Lane says I've got to deliver this merchandise personally. If I don't, he has vowed to deliver his size thirteen combat boots to my hind parts."

"Does this mean that Blevens, the Great City Man nonpareil, is actually going to Can Tho?"

"You heard me." I had thought he seemed kind of glum. Now I knew he was really put out.

"Blevens, ain't you too old for babysitting?"

"Wait 'til word gets out that Sergeant Blevens is nursemaid to a second lieutenant."

They went on. They were having a good time giving him the business, and he was taking it with relish, as I was the actual fall guy; and, weather-wise, I was starting to get hot.

"Is there an officer-in-charge here?" I asked.

"Sure there is. The Captain is usually found behind that door where the air-conditioning is." I started for the door. "But he's out, and the door stays locked."

"Sergeant Blevens," I'm trying to sound real natural, "what's our ETD?"

"Our expected time of departure is up to these pros here," letting me know who really runs show. "What do you say, guys? When are we out of here?"

"We got a special for you today. Helicopter. You rendezvous on the other side of the runway in about fifteen minutes."

"That's great. Thanks fellas. See you before the end of the week. Lieutenant Dinan, let's grab our gear and move out."

I told the sergeants what a joy it had been to meet them, gave them the usual unearned and unappreciated thanks, and followed my Sergeant. He picked up his gear, a small hand bag, and I picked up mine, two huge duffels and a footlocker. He made no move to help me, and I was not really sure about the protocol for this, but I was strong as an ox and figured I'd be just fine. How far could "the other side of the runway" be?

As we approached the runway, Sergeant Blevens said, "Gee, that's a lot of stuff you're carrying. I'd offer to help but it's sort of an in-country, unwritten rule that a man should be able to carry his own gear. Of course, if you can't handle it . . . ?"

"I can handle it, Sergeant." I hoped. Then, remembering the elevator zinger, I was afraid I'd been had yet again. Determined, I pressed on. This "across the runway" was becoming more like across the Sahara. I was bleeding sweat; my arms were wrenched. Now I was reduced to taking little steps; I was not going to make it! I stopped.

"You okay, Sir?" There was that little light in Blevens's eye.

"Sergeant, are you having me for breakfast?"

"No, Sir." Silence.

"Well then?"

"Sir, I'm having you for lunch, Sir."

I wanted so much to be mad as hell, but I couldn't help myself. I'd been so beautifully had, it was hilarious. I dropped my baggage on the edge of the runway and fell convulsed on top. I tried to speak. "Can you—ha, ha—can you, ha, ha, ha, ho, ho, oh, can you, ha, ha, can you believe the sight we must have made—ha, ha, unwritten rule my ass. You son of a bitch, ha, ha—Sergeant, I hope you're enjoying your lunch as much as I am."

"Sir! I—ha, ha—I guess I did have you on, didn't I? Ha, ha! Whatever can be sweeter than a second lieutenant for lunch? Ha, ha, delicious!"

"How about dessert?" I asked, reminding him of last Thursday evening.

"That was inspired—ha, ha, ha! You never mentioned how you made out. It's a standard gag under the circumstances. Some guys have actually browned their britches. You?"

"I came away clean, thank you, but just barely." It clicked. "You knew the driver didn't you? You called him by name, Nguyen."

"Naw, I didn't know the guy. Nguyen is the universal Vietnamese name. At least 90 percent are Nguyen something or other. It's like GIs being called Joe, a name you're pretty safe with. But I did tell him to stay off the main roads. I knew it would make your trip interesting."

"That it did, Sergeant. That it did. Abstractly, and just now, this is all very funny, but eventually I'm just likely to get personally pissed off. I suggest you keep finding ways of convincing me that I really need you. As of right now, you are at least two up on me, which can become a terrible bruise on my little ego."

"If we miss that chopper, Captain Lane's gonna put a terrible bruise on my little fanny. What say we move out? I'll take one of your bags."

"I hate to see you breaking the rules, but I think I'll get over it. Let's go!"

We waited while an F-4 jet fighter zoomed by, then hustled across the runway to some waiting helicopters. One was revving up. The flight sergeant, recognizing Blevens, was waving to us to hurry up. As soon as we were settled in, he took off. It's a different sensation, the first time up in a chopper. First it goes up, then it dips, then it goes up again. It's hard to explain—it's much more a feeling of being suspended in air than taking off in an airplane, but noisy! What a racket! Conversation was impossible. It was a thrill. Once up off the ground, the air wasn't so oppressively hot, and that was a true relief.

The sky was clear, I could see forever: the cultivated squares of gardens and rice paddies, the winding canals and twisting roads, a lone hut here and a grouping of huts there, all punctuated by specks of black smoke. The scenery didn't change, mile after mile; all cultivated, all touched by the hands of man. There were no virgin plains or forests. This property was in use. I still had an edge of excitement but was heading toward boredom due to repetition when we came upon a huge river of the brownest water I'd ever seen. Along its banks were lots of trees with some very tall and thick foliage as well as houses, roads, and paths dotted all along on both sides. But the real action was obviously on the water: hundreds of small craft of every description moving upriver, downriver, and across river, moving slowly with dignity, usually powered by a single oversized oar. On the water were also faster moving military power boats, leaving wakes behind. From fifty feet up one could see that they really didn't belong. On the left and right banks up ahead, the buildings started to bunch up. We were approaching a pretty good-sized city. I could make out the spires of what must've been a church and a couple of tall buildings with roof gardens as well as a marketplace, a bus station, a gas station, and then, suddenly, we were over an airfield. This must be our destination. It was. Down we went to a gentle landing. I felt no desire to kiss the ground, but I did feel a surge of excitement, of adventure. I said to myself, *Welcome to Can Tho, Lieutenant Dinan. This is going to be one hell of a year.*

FOUR

ARRIVING IN CAN THO

AS we stepped down from the helicopter, I was immediately confronted by a young man with private stripes on his sleeves, standing at what I had to believe was supposed to be attention. He was actually holding a salute in the very best military manner. I had almost forgotten about salutes (I would be reminded later on), but well-trained, I returned his salute automatically and said, "Good morning, Private."

"Good morning, Sir. Private Pribble, Sir. I'm here to escort you to Eakin Compound, Sir!"

I thought, where did this kid become a private? West Point? At least and at last, someone is showing me a little respect. "That's fine," I replied. "This is Sergeant Blevens. He'll be coming with us. Are we riding or walking?"

"I have a jeep, Sir. It's really too far to walk. Nice to meet you, Sergeant Blevens; I've heard all about you. You're something of a legend here—the ghost sergeant. Some people are convinced that you really don't even exist."

"I exist all right. You let them know that I'm the guy that's busting his ass every day in Saigon so that they can eat good in this here vacation spot."

The Sergeant was a little put off, I noticed. Evidently a sore spot had been touched. I'll have to remember that, I thought. Pribble noticed too. "I didn't mean nothing negative, Sarge. I'm just proud to meet you."

"Well, some people think I'm having a picnic up there" (I was certainly one of those people), "and that kind of thinking is bad for a man's rep. No harm done. It's nice to meet you too."

"The jeep is right over there, Lieutenant Dinan. If you'll show me your gear, I'll load her up."

"Thank you, Private Pribble. It's just this stuff right here. Come on, I'll give you a hand."

"No way, Sir! This is an EM job."

"Okay," I answered and began walking toward the jeep. Blevens was right alongside. "Say, Sergeant, aren't you going to help the young man?"

"Hell no, Sir. An E-7 simply does not *help* an E-3. Even a second lieutenant should know that."

I still hadn't formally, or informally, for that matter, acknowledged Sergeant Blevens's rank. The fact that he made a point of the E-7 rank told me that it bothered him to think that I looked upon him as only an E-6—great! After all, he was still up two big ones. Now was the time to let him know that I was aware of his rank all along and maybe score a point.

"I'm relieved to hear you say that, Sergeant. When I noticed that your sleeves didn't match your ID, I was afraid that I would have to ask Captain Lane to bring you up on charges."

"You mean that after all I've done for you, you would have gone and had me brought up on charges? Just like that, without saying anything to me?"

"That's right, Sergeant. It would be my duty." They say it's a good idea to get the word out that you're willing to heartlessly cut someone's balls off if that's what your duty calls for. It's supposed to make your men think twice before doing something really stupid. This was my first opportunity; I couldn't let it pass.

"When it comes down to it, you officers are all alike!"

"And you enlisted men?"

"Well that's . . . well, that's true in a way, Sir, but what I mean is, I guess we can't really be friends, can we?"

I had felt that there was that feeling moving around in Sergeant Blevens's mind, and I knew I was lucky to have it come up this way so that I could deal with it before it became a problem. "That's correct, whether you or I like it or not. That is correct."

Pribble was trying to pretend that lugging my baggage was no real hardship, but I knew better. That stuff was heavy, and in a few moments he was going to be doing the baby step shuffle, so I turned my back to him to preclude my observation of his humiliation. Looking around, I didn't see much to impress me. Actually the place was a dump. It literally screamed *temporary.* The Corps of Engineers could disassemble the whole place in a day. The runway was made up of metal sheets that hooked together like some sort of giant Lego set. Everything else was tents and sandbags. Upon closer inspection, I added another word to temporary: *boonies.* Yes, I was definitely in the boondocks.

The ride from the airfield to Eakin Compound was something of an anthropological extravaganza. The road we traveled on followed alongside a waterway that was about five or six feet across and probably four feet deep. You could tell that the water was moving, but slowly. This canal of sorts was fed by the mighty Bassac River. It seemed actually to be a plumbing estuary. This was the water supply: the bathing facility, the laundry, the kitchen sink, and the water closet for the families occupying the huts along its banks. There seemed to be the belief that the water was somehow self-purifying, and therefore everyone was having a grand time bathing, washing, or whatever, with no concern for their position as to up or downstream and no inkling of the many varied and lethal bacteria deposited by their upstream neighbors. So much for rustic beauty and the bliss of ignorance. This domestic lunacy was turning my stomach.

Naturally, my distress did not go unnoticed or unappreciated by Sergeant Blevens. "Say, Private, how about you pull over here so the Lieutenant can have himself a nice throw-up."

When the jeep stopped, the temperature leaped twenty degrees in an instant, and the smells that had been flying by as only a small irritation descended upon us like a net. I knew I wasn't going to be sick, but I wasn't sure that I could stand to breathe this air. Blevens never looked healthier or happier.

Thankfully, Pribble bailed me out: "You can't believe it now, Sir, but you'll get accustomed to it. I'm here only two weeks. At first I thought 'Hey, no way,' but now it doesn't bother me much at all."

"You're right about that, Pribble. I do not believe that I can ever get used to all this. Do you see the way these people are living? Why can't someone give them a few pointers on sanitation?"

Blevens replied, "All the sanitation they know, they learned from the French, and you know the inherent French disdain for cleanliness. We try to tell them, but so far they still have more respect for the French than for us. Their feeling is that if the French didn't bother about it, it must not be very important."

"Viva la France! Let's get going."

As we approached the city of Can Tho, the buildings became more substantial. You had the sense that there was real wealth here, not comparable to Saigon, but real wealth nonetheless. There were trees and bushes, flowers, gardens, gates, sidewalks, TV antennas, restaurants, hotels, shops, traffic (no traffic lights), gas stations—Texaco, Esso, Shell, a bus depot, a Kodak Photo store, dogs, children, people, the whole enchilada: civilization. Right away I liked it better than Saigon. The traffic was much slower; maybe it was the ox carts that kept the pace down. You could see that there were the rich and the poor, but you didn't sense the same disparity between the two that existed in Saigon. I was excited all over again. This was where the real people lived. This was where the Vietnamese were in control. We were only advisers, observers, here. This was Asia run by Asians. This was adventure.

We had passed through the main part of town, past the MACV IV Headquarters on the right and were heading to the outskirts on the other side when an arched gateway on the left announced Eakin Compound. We turned in under the arch. On our right was a walled-in soccer field; on the left, a facility that looked like a border-crossing shed from a grade B movie, then a warehouse, then a barracks for the Vietnamese guards. The warehouse,

I was to learn, belonged to the mess association and the "border-crossing shed" was just that. It was the checkpoint for the Vietnamese. They were searched arriving and departing the compound by their own Vietnamese police who used the guard barracks as a place to hang around, catch a nap, eat, or whatever. Directly in front of us was a very imposing looking machine gun emplacement. Making a right-hand turn, we traveled along the long wall of the soccer field. On our left was the rear side of what had to be living quarters, and up in front of us at the end of the wall, a yellow and green, stucco and wood-porched structure that announced: HEADQUARTERS EAKIN COMPOUND. Making a left, we passed by that building and headed down the right side of a pretty good-sized parking lot. We stopped in front of a long, single-story building, also yellow and green, which turned out to be the compound's colors. The roof of this building overhung the sidewalk like an Old West movie set. There were hanging signs identifying what lay behind each door. To the right, near the end of the building, a large sign read CAN THO MESS ASSOCIATION.

"Lieutenant Dinan," said Private Pribble, "Captain Lane instructed me to bring you right up to him. He wants to take you up to Compound Headquarters himself."

"Okay, let's go. Let's get this show on the road."

The screen door opened toward me, so we had to step around it to get into the office. The office was well lighted and poorly—or rather, not at all—decorated. One large room contained everything, including a cashier's cage that formed the rest of the space into a big *U*. It appeared unlikely that any two pieces of furniture shared a common heritage. The sounds heard came from adding machines, typewriters, and fans. In front of me, past a couple of desks occupied by attractive young Vietnamese ladies—each wearing a colorful, smart-looking, traditional Vietnamese dress known as an *ao dai*—sat a large redheaded American in jungle fatigues. The nameplate identified him as Captain Lane, QMC. I really felt unsure of what to do next, so resorting to my military training, I came to attention in front of his desk, saluted, and recited the usual formula: "Lieutenant Dinan reporting, Sir."

The Captain returned my salute, then stood up, extended his right hand in welcome, and also recited the usual formula: "Welcome aboard, Lieutenant."

This man was not large. This man was mammoth, a veritable redheaded, crew-cut, clean-shaven Goliath, six foot six, with large bones and a lot of meat. When he took my hand in his, I had the distinct feeling of being a child again, respectfully looking up at, and to, an adult.

The only sound heard came from the fans. Everything and everybody else had stopped. I could feel all eyes in the office burning into my back. My crew-to-be was sizing up their new skipper, and under the circumstances I figured I could expect to be rated a pip-squeak. I was outranked, out-aged, and out-sized. This guy would easily displace twice as much water as I would. "Big boots to fill," Sergeant Blevens had said. I'll kill him. He should have prepared me better. I was, it seemed in the overall, overwhelmingly outclassed.

"May I have your attention, please?" Could he have their attention? He couldn't lose it if he tried. "This is Lieutenant Dinan. He has been sent over here to be my replacement. I'm confident that with all your help, he'll do just fine." Somehow, it sounded like a put-down to me. This was not a cheerful welcome. The atmosphere was plainly funereal. The Vietnamese, now, all looked so sad. Was it his going or my coming? I hoped the former. Captain Lane came around his desk and stood beside me. He seemed, if possible, even bigger. "Come over here and let me introduce you." No one wanted to be first, so he started with the unfortunate who happened to be closest. "Shun, come on. This is Co Shun."

She stepped forward, giving me a small smile with a small nod of her head and a soft spoken "*Mon joi*"—and then did her best to make herself disappear, which was tough, as there was no place to go. This ceremony was repeated with Co Tuie, Co Hong, Co Nee, and Co Fu. An elderly man was introduced as Ong Trung. He actually came forward of his own accord, shook hands, and gave what must have been both a funny and embarrassing address. At any rate, all the ladies looked embarrassed, giggled, and turned their eyes away.

Captain Lane continued. "Hein had to go out, so you won't meet him until tomorrow. This is Sergeant Sullivan. He's been in the army more years than you've been alive, so listen to him." What? Another put-down?

Sergeant Sullivan came right over and took my hand as I extended it. He immediately struck me as a good guy. "Don't you worry, Lieutenant Dinan. You and I will get along just fine."

"Thank you, Sergeant Sullivan. It's nice to meet you. It's nice to meet you all, all of you." I tried with little success to make eye contact with each member of my staff.

"Well now," said Captain Lane, "you've met Pribble and Blevens here, so that brings us up to date so far. Why don't you wash up, and then I'll introduce you to the compound commander, and you can start getting squared away." This was clearly not a question. "Pribble, you'll show the Lieutenant to the latrine."

"Yes, Sir, Captain Lane. Please come with me, Lieutenant Dinan."

Out the door and to the right, first passing the Red Cross office, then the Post Office, continuing smartly, then wham-o, I almost ran into the latrine door that Pribble had opened for me. There was no getting around it; either I turned left into the parking lot or turned right into the can. Dutifully, I went to the right.

Pribble remained, so I suggested, "Private Pribble, I'm sure I can handle this all by myself. You may go back to the office. I'll find my way."

"Captain Lane won't like it if you get lost."

"Have a little faith, Private. I really will be able to make it all the way back on my own."

"Yes, Sir."

With this, he departed. It was good to be alone for a spell, even in a toilet. I tried to figure what sort of dope they took me for. Good grief, I am in for some fun—what's next? I had a sense of what the biblical David, armed with his slingshot, must have experienced coming up against his Philistine. But they were supposed to be irreconcilable enemies; we were supposed to be on the same team. Hell, we *are* the same team. Lane was passing the baton on to me, but it was there—hostility, directed at me. Nothing you could finger, nothing you could really specifically identify, but there nonetheless. What to do? Look, be cool. This one is definitely not pleased by what has been sent as his replacement. Too bad—he'll shove off as soon as he can, and I'll have to pick up the stick and do the best I know how. Meanwhile, lay low and learn. Not a very exalted plan of action, but it was something. It made me feel directed.

Freshly washed and combed, I returned to the office. As I opened the door, Captain Lane was already rising. "That took you a long time," he admonished. "Major Hicks called, looking for us. Take that thing off and give it to Pribble for safekeeping. You don't want to go walking into the compound commander's office looking like Billy the Kid."

"Absolutely not," I answered, thinking big boy here has a real case. I unhooked my pistol belt and gave the weapon over to Private Pribble. "Thank you, Private. And thank you, Captain Lane; that thing is really uncomfortable." It's always good to be thankful when someone's trying to be lousy.

"Let's go."

Back out into the sun. On the left, we passed the chapel-cum-movie theater; a little farther along, an arcade with the entrance to the EM club on the left and the entrance to the EM mess on the right; then four CONEX refrigeration units, the entrance to the officers mess; and directly in front I recognized the porch of the headquarters building. We were moving fast. The way his strides were outdistancing mine only served to reinforce my feeling of being a child tagging along after an adult. We passed several people on this short journey. Daddy got greetings; baby got stares. Oh well.

As we entered headquarters, the First Sergeant came right to his feet: "Captain Lane, you may go right in, Sir. The Major is expecting you."

"Thank you, Sergeant. This is Lieutenant Dinan, the one that's been sent to replace me." Ouch, the way he said that! Captain Lane was heading toward a door on the far left-hand side of the room and I naturally followed right along.

Someone loudly said, "Lieutenant Dinan." It was the First Sergeant to whom I hadn't exactly been introduced. "You wait out here, Sir. I'll tell you when you are called." Another little humiliation.

There was a Spec. 4 sitting at a desk on the other side of the office. In noticing him for the first time, I also noticed that he had the look of a satisfied audience. This performance had been anticipated and appreciated. I wanted to slap him and the First Sergeant, but I reminded myself: lay low and learn. There were a number of available seats, and something told me it wouldn't matter which one I availed myself of, any choice of mine would be ruled inappropriate by the Sergeant. The Specialist was the giveaway. He had that exquisite look of anticipation, like the school boy who has placed a thumbtack on the teacher's chair.

"Thank you, Sergeant. I'd prefer to stand if it's not going to be too very long, but just in case it does drag on, would you be so kind as to select a place for me to sit that's most convenient for you?"

As he turned toward me I caught a good-natured look that said: *You won that round. This is for fun, not meanness.*

"Help yourself, Lieutenant Dinan, to any chair but mine. That wasn't much of an introduction. I'm Sergeant Burreti, the Top Sergeant here." I took a step toward him, extending my hand. As he took it, he glanced over his shoulder to the Specialist 4th Class and said, "On your feet, soldier. Come over here and welcome Lieutenant Dinan. This is Specialist Four Grady, our company clerk. He doesn't look like much, but he gets the job done. Isn't that right, Grady?"

"You're always right, Top."

There was something of an inside understanding obvious here, as the kid was probably the best-looking Spec. 4 in the army. As we were all informing each other of our joy at meeting, the phone rang, and Sergeant Burreti sent me into the Major. I approached the door Captain Lane had gone through, thinking to myself, what now? I caused three distinct knocks: knock, knock, knock—and the response was, "Come on in." It actually sounded friendly.

As I opened the door, the coolness hit me square on—this office was air-conditioned! So I stepped in, quickly shutting the door securely behind me before really looking into the office. Turning around, I squared myself with the desk behind which sat he who was evidently The Boss. Looking very relaxed and very cool behind a large and uncluttered desk, sat a distinguished-looking Major clad in highly starched jungle fatigues, black hair slicked back with not one strand daring to be out of place. He was leaning back comfortably in a large executive chair, holding an elegant thin cigar between the index and middle fingers of his left hand, and smiling benignly in my direction.

I approached in a military manner and came to attention. As I started my salute, the Major came up out of his chair like a cat and thrust his hand toward me, greeting me with such real warmth that I was completely disarmed, "Hi there, Lieutenant Dinan. I'm Major Hicks, the compound CO. It's really great to have you here with us."

"Thank you, Sir. It's good to be here." And for the first time, I was thinking maybe it was.

"Sorry to have kept you waiting, but I needed to go over some details with Captain Lane." Captain Lane was seated in a chair to my left. He was looking big and unhappy. "I'll give Lieutenant Dinan back to you shortly. Right now, I want to give him my briefing, so why don't you go ahead and tidy up some of those details?"

"Very good, Sir, and thanks again."

I had no idea what he was being thanked for, but the thanks were sincere. They shook hands, and Captain Lane took his leave without a word to me.

"Well now, Lieutenant Dinan, why don't you get yourself a little relaxed," suggested Major Hicks, gesturing toward the just vacated chair while settling himself in behind his desk. I was happy to sit down, especially in this air-conditioned comfort. The thought hit me that although I had an instant liking for the Major, he did remind me of a used car salesman.

Settled in, Major Hicks continued, "Before I begin my pitch, I always like to give my new people the opportunity to ask one question, any question. It's the only free one you're likely to get—no strings, so I want you to think about it for a minute. Don't waste this opportunity."

With this, Major Hicks turned away and busied himself with some papers.

In a very real sense, I was left completely alone. Any question? But only one question? This is really not fair! How am I supposed to come up with one question that isn't going to make me look like a jerk? I mean maybe this one is for free, no strings, but isn't the question always going to be there? As in, why did he ask me that question instead of such and such. I know I can't really win in this, so I might as well buckle down and ask one that I really want answered but wouldn't normally be able to ask. I was watching the Major while trying to formulate my question.

Suddenly, he turned directly toward me and with great penetration looked deep into my eyes and said, "All right, let me have it."

"Why does Captain Lane hate me?" It came out of my mouth without ever passing through my brain. I couldn't believe I'd said it. I tried to back up, "I mean, well what I mean is, it's a . . . well, it's sort of . . . I think, a . . ." All the while, I knew I was sounding more and more stupid. You know, that helpless sort of stupid.

The Major never smiled, never frowned. He just looked at me benignly and took a thoughtful drag on his elegant cigar, letting the smoke out slowly, thoughtfully, wistfully.

When I finally had the good sense to shut up, he studied me for a moment more, then smiled broadly and exclaimed, "That, Lieutenant, is one great question. Do you know why that is a great question? I'll tell you why. Because it is questions like that which give me an opportunity, yes I said an opportunity, to explore and thereby to learn. And after all, isn't that what life is all about? Learning and sharing. Oh, I do like that question. Why does Captain Lane hate you? Let me ask you, does Captain Lane know you?"

"No."

"Then he can't hate you."

"I didn't mean to . . ."

"Don't get defensive on me, Lieutenant Dinan. Remember, this one is on the house."

"Yes, Sir."

"So, he doesn't hate you, but obviously you sense hostility. That means there must be a natural resentment—is that right?"

"Well, yes, I guess."

"*You* guess; *I* know. So now I will give you some of the facts of life."

"Thank you, Sir."

"Don't thank me just yet. Captain Lane has been around in this man's army for some time now. Everyone feels that he's done an outstanding job as the custodian of the Can Tho Mess Association. As a matter of fact, he has received the Silver Star in recognition of that outstanding service. Furthermore, when the Captain returns stateside, he's going to be promoted to major and given an assignment with even more responsibility. Now, Captain Lane is a big guy with a big heart, and he wants to like everyone, but when he sees you, a brand-new second lieutenant, standing there as his replacement, he's insulted. After all, how can a little pip-squeak second lieutenant replace him in all his glory? He knows people look at him and think, 'And you thought you were important. This is what the Army really thinks.' So of course he is bound to resent you. But I'll let you in on a little secret, the Army didn't pick you—I did."

Major Hicks settled back and helped himself to another long pull on his cigar, happily watching my perplexed state.

"You see, my real business is recruiting." I was stunned, a used car salesman in the flesh. "Recruiting requires real judgment, and I've rarely been proven wrong. This is a tough kind of assignment, and for my money, you have the best background for the job of all the candidates I was given. I thought old Danny-boy was pulling my chain when I first picked up your folder, but that fox is right out of recruiting too, and he has a great eye. Don't worry. You'll get over your rank. Every second lieutenant has to."

Major Hicks went on to explain what he expected from me, and how he expected me to behave, and how I could expect his full backing and all that sort of stuff. The Major had four months left on his tour of duty, and during all that time, I never saw him other than perfectly groomed, perfectly relaxed, perfectly in control, and perfectly happy to tell recruiting tales to anyone willing to listen. Actually, they were good tales. My Compound Commander, Major Marvin R. Hicks, was definitely one of Uncle Sam's top used car salesmen. He had an easy way about him and a keen sense of just how to handle any situation that came his way. The Sergeant Jones story is a great illustration of his insight.

FIVE

THE SERGEANT JONES STORY

SERGEANT Jones was an E-8, Army Corps of Engineers. With well over twenty years in the service, he was definitely a very senior NCO. It seems that the Sergeant developed a deeply spiritual and intellectual relationship with one of the local female residents early in his tour of duty, so that now, with only one month left on a twelve-month tour, our Sergeant was the reluctant father of a six-week-old little boy. Evidently, when Sergeant Jones handed over some cold cash by way of child support to the unlucky local, he let slip the fact that this was to be the last down payment toward the welfare of the unwanted because he was taking the big bird out of here before the Eagle's next landing. Being no Cio-Cio-San (Madame Butterfly), the demure did some fast thinking and, with the help of assorted sages, figured that a consultation with the local representative of the Great White Father in Washington might help to relieve the not-so-great Black Father here in Can Tho of some moolah on a timely and regular basis. With this in mind, the new mother gained an audience with Major Hicks.

There she presented herself, her story, and her child. Now, we all know that paternity can be a pretty tough thing to prove; it was almost impossible in 1967. But in this particular case, the evidence was overwhelming. You see, Sergeant Jones looked like Sergeant Jones and none other. When you encountered Sergeant Jones's face, you remembered Sergeant Jones's face. This was a face that only a truly big man was entitled to: huge eyes, huge lips, huge nose, huge jaw. It worked, it went together, it was unique, massive, and powerful. And this little one was the carbon copy made flesh.

Asking the evidence to wait in the anteroom, Major Hicks summoned Sergeant Jones for a consultation of his own. Now, when an enlisted man is summoned to the office of the compound commander, there is no question in his mind that something is amiss. A bolt from the place of evil strikes, instantaneously crippling head, heart, and bowels. The list of possible causes is almost endless. If you were lucky, it could be something palatable like the sudden death of some far-flung relation you really didn't care about anyway. On the other hand, it could be the unthinkable, the unbearable, the death or very serious injury of someone in your immediate family. In between lay all other conceivable pardonable and unpardonable possibilities such as bouncing a check, failing to sign for your malaria pill, tardiness at work, or too much celebrating at the club. Anything. It was endless. Contemplating this endless list of his possible indiscretions while trying to suppress the thoughts of familial tragedy, Sergeant Jones arrived at headquarters having accomplished his version of turning white as a sheet. Thus the worse for wear, Jones was escorted into the cool serenity of Major Hicks's office. The Major, starched, combed, and cool as always, released a thin and thoughtful stream of smoke with deliberate slowness before turning his most serene gaze upon the Sergeant's shaken countenance. Observing the Sergeant's extended state of discomfort, the Major became hopeful. After all, fear is often a forerunner of truth and of a willingness to cooperate. These two things were essential to settling this civilian complaint. So Major Hicks took his time, letting the tension build to a perfect pitch before beginning a dialogue.

Then, militarily: "First Sergeant, I am ordering you to remain as a witness to these proceedings."

"Yes, Sir."

Sergeant Jones cast his eyes hopefully toward the Top Sergeant. Nothing.

"Sergeant Jones, we have a very serious situation here, and I'm certain that with your help we can clear it up easily. You do want to help, don't you?"

"Oh yes! Yes, Sir! You know me. I always want to be helpful."

"Good, that's great. I told the First Sergeant here that you would want to be helpful. Isn't that right, First Sergeant?"

"Yes, Sir. Exactly right."

"Now, let me see, oh yes. A lady, a local lady, stopped by here to see me, and she claimed that you are the father of her child and that you don't intend to support that child. What do you say to that, Sergeant Jones?"

Sergeant Jones was visibly cheered. This was nothing. These girls were always claiming that this guy or that guy was their baby's father, but Sergeant Jones had been around; he knew it was impossible to prove. The correct line of action was clear: lie.

"That's impossible, Sir! I can't believe that you would believe a lie like that. I have never looked at one of these slant eyes, much less made babies with them. I'd be afraid to, you know. Look at the size of me; I know I'd just crush one of these skinny little girls to death. No, Sir. Not to mention the diseases they carry. Besides, I'm a happily married man, and my wife wouldn't like me doing nothing like that—no, Sir!"

Major Hicks was thinking, *Damn, I thought I had him wound up just right. This guy's a pro—great recovery.*

Then, to the Sergeant: "I'm really happy to learn that from you, Sergeant Jones. I know that you would want to be helpful and that you wouldn't distort the truth (*never say lie*) in an important matter like this."

"That's right, Sir. That's right. Is there anything else or may I go now?"

"No, you're free to go unless, Sergeant Burreti, have you anything to add?"

"No, Sir."

"Very well, then. Oh, just one thing before you leave. Sergeant Burreti, would you bring our guest in?"

"Yes, Sir."

Sergeant Burreti stepped behind Sergeant Jones, crossed to the other side of the office, opened the door of the anteroom, and escorted mother and child back to the Major's desk. Sergeant Jones was decidedly looking the other way.

"I want you to take a look at this baby here, Sergeant Jones."

"Oh no, Sir. I don't want to look at no baby." Sergeant Jones had a real anti-desire to look at the child.

"Sergeant Jones, this is an order. I want you to look at this baby."

Sergeant Jones looked about, and in looking at the child, he knew he was looking in the mirror.

"Now, Sergeant Jones, I want you to tell me that you are not the father."

"Oh no, Sir. I mean, yes, Sir. I mean, oh yes, that's my child. I can't deny it. It's my child all right. Yes, it's my child."

"Thank you, Sergeant Jones."

"Miss, you can go home now. Sergeant Jones and I will work this out." When they left, Major Hicks invited Sergeant Jones to sit. The big man had tears in his eyes, and the Major's instincts told him that they were there for the right reasons. "Why don't you tell me about it, Sergeant?"

"My wife and I, we haven't ever been able to have a child. I always wanted a son, but now that I have one, it's not really mine because it's not supposed to be, and anyway, I'm leaving, and he's staying, so I figure it's best to forget about the whole thing."

"You cannot just forget the whole thing. There happens to be a human life involved here, and let me tell you, this kid won't stand much of a chance here. If you really want a son, you can have him."

"What are you saying?"

"I'm saying, why don't you take the boy home with you?"

"You're saying I can?"

"I'm saying you can."

"Good God, my wife will kill me, but then she'll forgive me. Are you sure?"

"I'm sure."

"How do I know that his mother will let me?"

"She'll let you."

"You sure?"

"I'm sure."

The American forces had been here long enough to learn from firsthand experience that the children of mixed parentage had a really rough go of it in Vietnam. The offspring of Negro-Asian liaisons were particularly looked down upon and regularly degraded. The mothers also could expect to be treated like lepers. It is a great credit to the values and courage of these mothers that infanticide was never a widely practiced solution to this long-range liability. Therefore, the Major was safe in his assurance that the mother would release the boy to his father. Naturally, face had to be saved, and that was accomplished in the time-honored method universally known as cash payment.

Major Hicks saw to it that all necessities were accomplished with record speed, and when Sergeant Jones left Eakin Compound, he left with the baby in his arms and an enormous smile on his face. He wrote back after being home for a while, and it seems that when his wife saw the baby, the only thing she could see was her husband. She said it was the happiest moment of her life; that now she understood how a mother feels when she first sees her own newborn, and because of that she would always have this child as her own.

Back to the present. Having finished his briefing, Major Hicks stood up. Naturally, I did likewise. Coming around from behind his desk, he escorted me to the door, letting me know, "Lieutenant Dinan, again, I want to tell you how glad I am to have you here. I just know that you'll do a fine job. Now I'm going to turn you over to Specialist Grady, who will show you where you're bunking for the time being. After you get your gear squared away, you can go back to the mess office."

"Thank you, Sir. You know I'll do my best," and I meant it too. This was a guy who made one want to shine.

Major Hicks opened the door and called to Grady, "Specialist, I want you to take over now and be sure to take good care of Lieutenant Dinan."

"Yes, Sir. Lieutenant Dinan, won't you please come with me. We'll get your stuff, and I'll show you to your quarters."

"Lead on."

SIX

THE MESS HALLS

WE went back out into the noonday sun and retraced the route I had taken with Captain Lane a little earlier. Now there was considerably more activity than before with everyone seeming to descend upon the entrances to the mess halls. Back at the jeep, we collected my belongings, and I followed, this time to the left, past the mail room, past the barber shop, past the snack bar, stopping at the last door on this corridor. Immediately on my left was a large grouping of trash cans; just ahead and to the right of them was the PX to which they belonged. Specialist Grady knocked; getting no reply, he unlocked and pulled open the door to my new home. This place needed a lot of help, but even with lots of help, not much could be done. It was a small room with two bunks, two dressers, one armoire, one ceiling fan, and one window. Out of the back window you could see what appeared to be a very bleak camp that turned out to be the hospital for wounded Vietnamese soldiers. Having no desire to spend any kind of time in this dump, I tossed my things on what was obviously the unused bunk and started out the door still held by Grady. Naturally, he was enjoying my evident disappointment.

"Great digs, huh, Lieutenant Dinan? I wish I had a place to live like this."

"You've been very kind in showing me here, Specialist. Am I supposed to tip you?"

"It's up to you, Sir. It's not required."

"If I liked the room better, you'd have a chance. For this, no way."

"I think you must be spoiled, Sir. I think this is a great room."

"I'll find a way to arrange for you to have it."

"You really don't have to go that far, Sir. Really."

I think the idea must have hit him that maybe I *could* do this to him; it backed him off a little. Actually, he seemed like a good guy.

"OK, I won't for now. Are you supposed to give me that key?"

"Oh yes, here it is, Sir. And Sir, remember to always lock things up. These folks claim that if you don't, you're clearly telling them that you really don't want whatever is there, thereby giving them permission to take it with a clean conscience."

"That's cute. I'll try to remember that. Thank you."

To give you an idea of just how undesirable this habitat was judged to be, it should be noted that this place was turned into a storage room when CWO Arnold shipped out and I moved into C-5. Even at the height of our billeting needs, it was never again given consideration as a place of dwelling.

Specialist Grady went about his business, and I returned to the mess office, which was deserted but for the person of Captain Lane, who was busy doing whatever at his desk. As I came in, he looked up and said, "There you are. Let's go get some lunch."

"Sounds good to me."

I stepped back outside and held the door for Captain Lane, who was right on my heels. He locked the door carefully before we headed off. Leading up to the entrance of the officers mess was an enclosed walkway. The walls were studded with clothing hooks; during the rainy season this arrangement was a godsend.

Just inside the door we were greeted by one of the mess personnel who was handing out little order forms in exchange for either the showing of a current mess association identification card or fifty cents. Captain Lane introduced me and vouched for the fact that he would make sure that I paid at the office. I followed Lane to a table on the far wall of the dining room. From here, we could observe the whole room. I liked it immediately. It had a good feel to it and the right kind of smell. There was no air-conditioning, but between the ceiling fans and the filtered light, there was a coolness to the place. The floor was green tile. The walls were white wainscoted up to chair rail height and light blue above that, as was the ceiling. The furniture was great-looking stuff—bamboo and rattan. The backrests and seats of the chairs were padded in soft orange leather. The rectangular room was regularly set with tables for four in neat military rows. Off to the far left-hand side of the room was an alcove, semi-enclosed with a table for twelve. This was the generals table.

Looking back toward where we had entered, past the tables, was a large buffet laden with self-help items such as salads, bread, butter, sliced fruits, and desserts. Each table had a number on it, and the system for service was very simple. When you picked your table, you noted its number on your order slip and checked off your choice of entrée. These slips were then picked up by one of the white-uniformed Vietnamese waitresses who would serve you when your order was prepared. Altogether, a very civilized operation.

Today's choices were Southern fried chicken, lamb stew Parisian, and roast beef sandwiches. I ordered the lamb; Captain Lane, the chicken. Then we made ourselves salads of local lettuce and tomatoes. I opted for the French dressing; Lane, the Roquefort. We helped ourselves to some French bread and returned to our table. The salad was fine, the bread excellent, and when our main courses arrived I was pleasantly surprised to see how inviting everything looked. My stew was delicious, and future experience would assure me that the chicken was too. For dessert, I was encouraged to try the local pineapple. Happily I did. For me, Vietnamese pineapple is the best in the world. Over coffee, Captain Lane outlined his plans for me for the next few days. Today, he was anxious to get me signed up as a member in good standing of the mess association, after which I was free to explore the compound, thereby familiarizing myself with my surroundings, and to organize my equipment, thereby being the more ready to perform my duties. As advisers, we found it politic to follow the Vietnamese custom of the two-hour lunch—definitely my kind of war. Therefore, since the office would not reopen until 1400 hours, he suggested I might wish to take the opportunity to observe the operation. I said I would really like to, but could I do it from inside the kitchen? This seemed to shock him.

"I don't know why you would want to watch what actually happens inside the kitchen, but listen, if that's what you want to do, remember you're going to be the big boss here in a few days and every door has got to be open to you."

"That's great. You know, I'd just like to get a little feel for the guts of the operation. Everything seems to run very smoothly."

"That it does. Come on, I'll introduce you to Sergeant Chavez. He's the NCOIC (Non-commissioned Officer-in-Charge) of this facility."

Inside the kitchen, the activity belied the calm of the dining room. The kitchen, centrally located, was, for the most part, supervised by Americans and powered by Vietnamese. We stood by the service counter for the officers dining room. Fortunately, the cook neither knew nor cared in which direction the finished product was going, so the EM mess and the officers mess got equal treatment in the food department. The enlisted mess was about two and a half times the size of the officers mess, and although it was very nice in general terms, as it was larger, it was noisier, and the quality of the décor was a couple of notches down.

Sergeant Chavez was busy supervising the activity, much as a chef would be in any good kitchen, and his displeasure at the choice of time for an introduction was evident. "It's very nice to meet you I'm sure, but right now, as you see, I'm very busy."

"That's all right, Sergeant, I understand. I've been a professional cook. I'll just stay over here out of the way and watch, if that's okay. By the way, the lamb was delicious."

Having made his introductions and seeing an exchange in progress, Captain Lane withdrew with an "I'll see you at 1400 hours." He didn't wait for an answer.

"You were a cook?" Sergeant Chavez inquired.

"Yes."

"Where were you a cook?"

"In New York City, at a restaurant called The '21' Club."

"You're kidding!"

"No."

"I've heard about that place; we'll talk later. For now, the end of the counter over there is the best place for you to observe."

I went to the end of the counter and was happy to watch this kitchen working like a well-oiled machine. These people obviously knew what they were doing. Everyone showed great deference and respect to the Sergeant and he, in turn, seemed to be genuinely happy in what he was doing and with those who were helping him do it. As things started to perceptibly slow down, Sergeant Chavez turned the kitchen over to one of his subordinates and invited me to join him at what was evidently the chef's table. Every kitchen that can afford the space, and even some that can't, has a chef's table. To be invited to it, is no small honor. When we sat, he removed his hat revealing a full head of thick, wavy black hair streaked with gray.

He smiled a big, warm, Mediterranean smile showing strong, even, white teeth and said, "Lieutenant Dinan, I hope you can make good on your claim of being a professional cook, because you're the first officer to be invited to sit at this table and you're invited as a fellow cook. If it were to be learned that you were not a cook, I would lose face and regaining it would be a serious matter."

In a way I felt I should protect my position as an officer to an enlisted man, but something in this man's dignity and sincerity made it more important to me to defend my honor as a cook of merit.

So I responded, "Sergeant Chavez, I want you to know that from what I have observed, I think you are running a very fine kitchen here, and I am therefore very honored by the invitation to join you at the chef's table. There is one other thing I must say, and that is that I will never tell a lie to you or to anyone else under my command. It's important to me that you understand that."

"Please understand, I was not suggesting that you were lying, Sir, but most people don't take the food business very seriously. If someone grills hot dogs for two weeks at summer camp, he's likely to tell everyone that he's a chef."

"You're right. You're absolutely right, Sergeant. I withdraw any suggestion of offense. You know, when I was in the Intelligence Corps, before going to OCS, I did a stint working for DoDNACC (Department of Defense National Agency Check Center). We processed security check files. It was real boring work, so we would read the ones that looked interesting for a laugh. It seemed that every time we came across a real lulu—you know, a rapist, sodomist, arsonist, burglar, mugger-cum-general malcontent—some enlightened officer would recommend this sterling individual to the Food Service Industries. It still makes me mad as hell."

"We don't always get the cream of the crop in the kitchen. I'm happy to learn that you're commissioned from OCS. You guys are the best for my money. How did you get involved in cooking? I take it, it was not on the recommendation of some enlightened officer."

"No. I was studying Hotel Administration, and food classes were required in the first semester. I found I enjoyed the food classes so much that I elected to take the Culinary Option in my following semesters. During the summer break, we had to work in the industry. Through my cousin Lois's connection with Jerry Berns, one of the owners of '21,' and the recommendation of the school's president, 'Mama' Lefler, I was offered the apprentice job at The '21' Club. The job was supposed to last ten weeks. At the end of that period, the Chef, Louis, suggested that I quit school and study with him. I compromised and worked part time, which pretty soon turned into full time, and following graduation, I stayed on and became Chef Décorateur, and later, Chef Poissonniére, and later, a private in the Army of the United States. In the meantime, I had worked on all the stations in the kitchen and Local #89 in New York City is currently holding an A card with my name on it."

"You could have continued your cooking in the army."

"I know, but I thought that while in the army I might as well try something else. Therefore, the Intelligence Corps. But at the time I was applying for OCS, my CO happened to be a foodophile, and he recommended me to the QMC. So here I am, back in the kitchen."

"I'm happy to have you in my kitchen, Sir. You're always welcome." The *my* was not lost on me.

"Thank you, Sergeant Chavez. I know that *your* kitchen will be a very positive part of *my* performance report." I did have to make sure that he knew that I knew that I was the only one ultimately responsible.

We made a little small talk, swapped a few preparation techniques, and pretty soon it was almost 1400 hours.

"I'd better be going; Captain Lane will be expecting me. Great meeting and chatting with you."

"Likewise."

I was off to see Captain Lane. Later, I would come to learn a whole lot about Sergeant Chavez. He was really quite a guy.

As I was to discover, Sergeant E-8 Chavez's obvious competence, his intimate working relationship with the Vietnamese employees, and his easy air of this-is-my-kitchen did not happen by some unusual quirk of fate or personality. You see, upon my arrival, Sergeant Chavez was already deep into his fourth year on the job at Eakin Compound. When he had arrived, it was only a small advisory outpost. He had grown with the operation, and the operation had grown with him. He was one of the people responsible, on the frontline, so to speak, for the way the operation actually functioned. It wasn't by accident that he seemed like a chef in his kitchen. In very real terms, that's exactly what he was.

SEVEN

SERGEANT CHAVEZ AND MY FIRST SHELLING EXPERIENCE

AS for the true cause of our good fortune: as usual, the gentle sex. In the course of Sergeant Chavez's first tour of duty, he fell in love with and married a local lady, and because of all the difficulty involved in having that marriage accepted by our authorities and taking her out of the country, he decided to keep re-upping.

In those years, before the Tet Offensive of 1968, it was a fairly gentle war here in the Delta. We would be reminded later on by the lyrics parodying the Beatles song *Yesterday:*

Yesterday,
War was such an easy game to play.
The VC seemed so far away,
Oh, I believe in yesterday.

So he stayed on and, I was to learn, lived pretty much a civilian life, keeping a domicile in town and commuting regularly on a daily basis from home to work and from work to home.

This, of course, was against all regulations, and although it was common knowledge to all the Vietnamese and a select group of enlisted personnel, it escaped the keen observation of the officer ranks until one particular night when the enemy decided that we needed to be reminded that there actually was an ongoing war in progress. On that fateful night, a barrage of mortars and rockets were launched at Eakin Compound. Most of what was intended for our benefit landed behind us in the Vietnamese Hospital compound. Some sank harmlessly into deep mud. Some landed about the compound but fortunately never exploded. One mortar, however, broke through the roof directly into Sergeant Chavez's room and exploded. The compound was in a state of upheaval. Many a nose was bloodied by doors flying open into the faces of those at cross-purposes between running down a corridor and those running out of a room. One officer was severely bloodied in a valiant struggle with some metal lawn furniture. Several enlisted men had to be pulled out of drainage ditches. All in all, we were shown to be ill-prepared to deal with this "reminder." When things calmed down, it was time to assess the damage. The worst part proved to be the horrible lamentations from the Vietnamese Hospital compound. Naturally, a lot of attention was given to the one place on Eakin Compound where real damage had been sustained: Sergeant Chavez's room. Those who made the search were happy not to find body and blood spattered about. The OIC, not knowing whose room this was and needing the information for his report, made diligent inquiry and asked that he who was fortunate enough not to be at the right place at the wrong time report immediately, so that an assessment of the damage to personal and government property could be made for his report.

At some point it became quite clear that Sergeant Chavez was not within the confines of the compound.

I was called to Major Hicks's office, and the Major, cool as always, questioned me: "Lieutenant, have you any idea where your Mess Sergeant could be?"

"No, Sir."

"Do you know that he is at this time, officially AWOL?"

"I hadn't thought of that yet, Sir. I've been too busy being glad that he wasn't found splattered all over the place."

"That was my first reaction also, but this is a very serious situation we have here. It was way past curfew when the fun started, and Sergeant Chavez is unquestionably Absent without Official Leave. The entire compound knows it, and my guess is that more than one guy out there is watching carefully to see what you do about it."

"I'll have to do my duty, Sir, and write it up as it is."

"Of course, we all must do our duty. But first, let us find out how it really is."

With this, Major Hicks lit up one of his elegant, slender cigars, leaned back, and took a thoughtful pull.

Slowly letting the smoke out, he then turned his attention back to me and said, "You know, the Sergeant has given this place almost four years of faithful service. Next month, he's scheduled to leave here and the Army. He is going to take his wife with him and rejoin the civilian world in Hawaii. As an E-8 with close to thirty years behind him, he should have a pretty good and very well-deserved pension to help him out. I don't think it would serve military justice very well if he were to have his record spoiled because he fell asleep this one time at a friend's house. Do you?"

"No, Sir."

"So, I'll tell you what to do, Lieutenant Dinan. You go and wait for Sergeant Chavez at the gate. When he comes in, you bring him to my office, and don't let anyone speak to him. You stay with him and send someone to get me, and I'll come and talk to him alone so that I can find out that that's exactly what happened."

"That sounds great to me, Sir."

"The important thing is that I get the right story for the record. Do you understand that?"

"Yes, I do."

"Good then, I'll go back to bed, and I'll see you later this morning."

It wouldn't be long before dawn started to break, so I just headed right on to the gate. I had no idea when Sergeant Chavez might show up, but it was clear that Major Hicks's compassion for a fallen human would not extend to me should I botch this mission. As it turned out, I waited about two hours, giving me plenty of time to consider and appreciate the Major's unique character. With all that confusion, he was, as always, perfectly groomed, perfectly relaxed, and perfectly in control, focusing on the key issue and finding the avenue for real justice.

Around 0700 hours, Sergeant Chavez came strolling through the gate. With him were a few of our Vietnamese employees. When he saw me he became so disoriented that he even saluted me. As a reflex I returned it and then waved him quickly over. By 0700 things are already pretty busy on a military installation, and for some reason, in this we were no exception.

I had my speech all prepared, and as soon as he was close enough that I could speak softly, I told him, "Don't ask me anything and don't speak to anyone more than to say hello. This is a direct order. Do you understand?"

"Yes, Sir, but . . ."

"No buts, Sergeant, I'll tell you again. Say nothing to anyone more than hello. This is a direct order. Do you understand?"

"Yes, Sir."

"Follow me."

He did. A direct order is one of the greatest of military traditions and tools. It cuts through all the usual bullshit when necessary. The giver, who must outrank the receiver, assumes all responsibility for the outcome of the order. The receiver is thereby given complete immunity. The consequences of disobeying a direct order, on the other hand, are catastrophic. I knew with certainty that even if the commanding general were to call upon Sergeant Chavez for more than a hello, he would not get it.

We went directly to headquarters. As we entered, Sergeant Burreti was instructing Specialist Grady, "Grady, I want you to go directly to Major Hicks's quarters and let him know that Lieutenant Dinan is waiting for him in his office."

"What about Sergeant . . . ?" He was quickly cut off.

"Grady, just do exactly as I've instructed you."

"OK, Top. I'm on my way."

Without a word, Sergeant Burreti led us into the Major's office, and immediately returned to his desk. One look at Chavez was sufficient for even the most unobservant to see that he was contemplating the worst. I tried to take the edge off: "Cheer up, Sergeant. The Major is on your side. When he gets here, I'll be going, and he'll speak with you alone. I think that, for the time being, it's best for both you and me to say nothing."

"Thank you, Sir. I take it that your direct order does not apply to Major Hicks."

"That is correct. Once he comes through the door, consider the order rescinded."

Pretty soon the Major arrived, fresh as a daisy, his usual self. We stood as he entered and remained so as he settled himself in behind his desk.

He addressed me first. "Thank you, Lieutenant Dinan. You may go about your business. Should anyone ask you what's happening here, you tell them that I am getting to the bottom of the matter—nothing more."

"Yes, Sir."

As I was on my way out the door, I overheard the Major telling Sergeant Chavez, "A lot of people got very lucky last night, maybe you more so than most. We want to be sure that it stays that way."

Evidently Sergeant Chavez confessed that he had, indeed, fallen asleep at a friend's house, and to be certain that this would never happen again, he agreed not to leave the compound for the remainder of his tour of duty. It was determined that, in this instance, no charges were necessary and that the Major's stern verbal reprimand was sufficient punishment for this one-time offense.

Because he was confined to post for his last month in-country, the big surprise party that our Vietnamese staff had planned to hold downtown had to be cancelled. At the party they were going to present Sergeant Chavez with a gold bracelet. The custom was that the jeweler would affix the bracelet to one's wrist so that the only way to remove it would be to cut it off. They got special permission to bring the jeweler into the compound, and he attached the bracelet to Chavez's left wrist right at the chef's table. The bracelet was 18-carat gold and heavy. To pay for such a gift, everyone had to have given until it hurt and hurt plenty. Sergeant Chavez was touched and embarrassed at the same time. In response to their generosity, Chavez made up gift packages of toiletries, items very hard for the Vietnamese to obtain, to give each and every worker. It was typical of Chavez that he insisted on giving back more than the bracelet had cost and when one is responding to 150 people, that's not very hard to do.

When he left at the end of those four weeks, there were a lot of genuine tears shed in sadness. For me, the saddest thing was that some of his peers continued to grumble in disappointment because they felt he had not been sufficiently disciplined for the crime of not being in his bed where and when he should have been, and where, not incidentally, he surely would have been killed on that fateful night.

EIGHT

EXPLORING THE COMPOUND

I arrived back at the mess office a little before 1400 hours. A couple of the Vietnamese women were waiting under their parasols while Private Pribble was meanwhile struggling with the lock. By the time we all got into the office, and Pribble and the young women settled, and the lights turned on along with the fans, the clock was showing 2:00—it must have been a civilian clock—and Captain Lane was coming very punctually through the door.

"Good, I'm glad you're here, Dinan. I want to get you signed up and paid up for this month's mess association dues. I always make a habit of being the first to pay every month. It's a custom you should take over."

Taking his seat just as the word *over* finished leaving his mouth, the Captain looked up at me with the look that said, "It's time for you to say something." I usually prefer to feign ignorance in such situations, but somehow it didn't seem appropriate to the occasion so I dutifully said, "It sounds like a very good idea, Sir."

"Yes, it is. Dues, which cover all services rendered by the mess association, are set at the full amount of your COLA. I've computed the amount due from you including today's lunch. You will give Private Pribble $47.50, and he will issue proper mess identification to you. The rest of the day is yours to do with as you please. I suggest squaring your belongings away and inspecting the compound to get a feel for the place. You will report to this office at

0800 hours tomorrow in combat fatigues and boots. Your weapon will not be necessary. Any questions?"

"No, Sir."

Captain Lane, with no further word, immersed himself in something on his desk. I was no longer there. So much charm, this one. You really have to like a guy like this. I paid my $47.50 and received my validated ID from Private First Class Pribble, who served as the cashier for the mess association. He also returned my .45 hand weapon, complete with holster and belt. I thanked him, and as he headed back into the aforementioned cashier's cage, I headed back to my quarters, figuring I might just as well follow the Captain's advice.

I had the notion that, on second thought, perhaps my room would be less depressing. I was wrong. So I arranged my stuff and put it away as quickly as I could, changed into my combat clothing, and headed out for a stroll about the compound.

I headed to the right, away from the direction of the mess office. I was out into the sunshine for just a couple of paces, then up three steps onto another covered walkway of basically the same construction as the one I'd left, but you could tell that this building was all billeting (military lodging facilities). On my left, I passed the backs of the PX, the engineer's office, and two small buildings: one locked up, the other open for business. Halfway along the walkway was the bathroom. Inside, I found the expected: four showers, four sinks, four toilets, and two urinals—and the unexpected: the place was really clean. The path I was following then led me past the rear of the officers club and back out into sunshine, then down three steps where it turned to the left and then to the right, up a short rise where, lo and behold, I was confronted by an enormous swimming pool filled with the bluest, coolest looking water imaginable. To my right was a large deck, not quite half-covered to protect the sensitive from the sun. There was the usual assortment of pool furniture: lounges, tables, chairs, and a barbecue. This entire area was enclosed by bamboo paneling that ran the length of the pool and around a smaller covered deck fitted with screens for changing. And to think, no one had even suggested that I pack a bathing suit.

Walking down the length of the pool, on my left was the back of what had to be another building devoted to billeting. Up in front I couldn't see beyond the bamboo paneling, so my next surprise had to wait until I cleared the pool area. Stepping past and to the right of the protecting wall in front of me, I beheld a tennis court. That's correct, a tennis court! And to think, no one had even suggested I pack my tennis whites.

Beyond the tennis court, I could see the back side of the machine gun emplacement, the front side of which I had seen on my way into the compound a few hours ago. This cheerful sight, along with the abundance of strategically placed, sandbagged bunkers scattered about the compound, served as a not-so-gentle reminder that we were also seriously engaging in the sport of war. This place really had something for everyone.

Passing the gun and crossing the driveway, I followed the sidewalk that ran along the long wall of the old soccer field heading toward the compound's headquarters building. To the right of headquarters, three steps up found me on yet another covered walkway, along which I encountered the compound supply facility, transient billeting quarters, and the compound security barracks. I was now on the enlisted side of the compound. This was the only barracks style facility in use at Eakin Compound. There were two unofficial classes of enlisted personnel in this compound, the *doers* and the *protectors*. The *doers* at the lower end of the ranking order (privates and specialists, E-1 to E-4) were billeted in two-man rooms. The *protectors*, who were all at the lower end, were provided with a twenty-man barracks. In fairness, they had to be kept together and watched carefully. These were the troops who, upon taking the various aptitude tests, were found to have none. Certainly the wisdom that decides that it requires more aptitude to be entrusted with a spatula than a gun might be considered suspect. But, "Ours is not to question why; ours is but to do or die."

At the end of the corridor, it was down three—sunshine—up three and turn left. I was now on senior NCO row. It was here that such notables as the

Sergeant Major of IV Corps and the compound's First Sergeant, along with the rest of the non-commissioned officer elite, were quartered in the comparative luxury of single occupancy rooms. (Commissioned officers gained this status at the senior major level.) Midway down on the left, the first door of the connecting billeting row was overhung with a sign proclaiming MARTHA RAYE LOUNGE. A smaller inscription on the door advised: ADMITTANCE TO SENIOR NCOS ONLY.

During this period of time, Martha Raye was roaming at will all over Vietnam. Half of the time, the authorities had no idea where she was. Anyone who could be even remotely held culpable should any evil befall her, always felt the thrill of relief when she showed up alive and well. She stopped at Eakin Compound several times during my stay there, and every time the same story would be repeated: dinner with the General and his staff was obligatory. The challenge was always to see how long they could detain her from joining the company of her choice, namely, the senior NCOs. I remember the first time I was introduced to her by Sergeant Burreti. I was walking across the parking lot as Sergeant Burreti and Martha Raye were coming from the opposite direction. From a distance, she appeared to be just another army person in combat boots and tropical fatigues, but as she came closer you could see she was also wearing that one-of-a kind smile.

As we neared each other, Sergeant Burreti spoke up: "Lieutenant Dinan, I'd like to introduce you to our favorite lady, Martha Raye. This is Lieutenant Dinan. He's in charge of all the clubs and messes and all that stuff."

"It's an honor to meet you, Miss Raye. It's really great to have you here. After all, our senior NCO club bears your name."

I knew at that point I must have sounded like a jerk, but what could I do; I didn't know what to say. *She* did: "Call me Maggie, and remember, I don't like boys and I don't like officers. You're both." The Sergeant was beaming. I was not.

I said with some venom, "A real pleasure, Maggie."

She replied in the most disarming way, flashing all those teeth, "Oh heck. I didn't mean to come down so hard on you. I'm sure you're a really nice kid.

But I just endured lunch with the General and the rest of those assho- (mid-word correction) officers. God, are they dull." To First Sergeant Burreti, she said, "Why is it that the officers become cheap and dull and the enlisted men become generous and interesting?"

The First Sergeant was smiling as he suggested, "Say, that should be a great topic for you and the fellows to chew on tonight."

I left with a "Thanks, Maggie. Please excuse me; I'm running late."

She flashed me her smile, "See you 'round, kid."

In the senior NCO club, aka the Martha Raye Lounge, was a very handsome portrait of the lady done by a senior NCO of no mean talent. It was inscribed, "Thanks to all my friends, Maggie."

Continuing along, on my right were more doors, and on my left more connecting rows coming in at right angles. At the end of the corridor, it was once again down three—sunshine—up three and left, down another covered walkway with doors on my right and left, some grass, and then the closest row building. Straight ahead, just to my left, was a towering concrete structure that turned out to be a four-walled handball court. And to think, no one even suggested I bring a pair of handball gloves. Just beyond the handball court was what I guessed to be one-half of a regulation-size basketball court (obviously requiring one sneaker and half a ball). And beyond that, directly in front of me, down three steps—sunshine—keeping on level, a scant distance ahead was the mess office again.

Instead of returning to the office, I took a left across the basketball court and followed a walkway that traversed the interior of the compound. On my right was the rear side of the chapel/theater, then the EM club. On my left was a fairly good-sized yard of CONEX containers. Off to my right, through a large breezeway, I could see the parking lot and the officers side of the compound on its far side. As I continued, I passed a building cheerfully marked CAN THO MESS ASSOCIATION PROPERTY—UNAUTHORIZED PERSONNEL KEEP OUT that lay to my left. Now the EM mess and the kitchen were on my right, and the laundry was on my left, its yard hung with sheets trying to blow in the nonexistent breeze. Next on my right was the officers mess/dining room followed by the rear of the headquarters building. I turned to look back at all I had passed and saw that behind the laundry and storage facilities were several rows of identical long, low buildings for enlisted billeting, all of which dead-ended into senior NCO row. Between the rows were grass, picnic tables, and many sandbagged bunkers. All in all, it seemed rather orderly and pleasant.

I walked between the officers mess and headquarters and crossed the parking lot to the officers side of the compound, which consisted of several more long, one-story parallel rows of quarters buildings. Turning right, I came to the first of these rows. Up three steps was a bamboo door that advised SENIOR STAFF ONLY—OTHERS KEEP OUT. I was *others*, and I did. To the right of this welcome sign, instead of the usual grass area between rows, was the general officers lounge, later to become the general officers mess. The next row was level with the walk, no three steps. Its front faced that of the third row with connecting walkways and lawn. Both dead-ended into what one rightly would figure to be senior officers row. The back of the third row shared lawn space with the back of the fourth. This fourth row was Row C, which was to be my home for most of my stay. I now faced first, the Post Exchange, then the Corps of Engineers building, and then the volleyball court. I was to learn that the net, set at eight feet for volleyball, was adjustable and could be reset at five feet, one inch for badminton. And to think—ah well, volleyball wasn't really my game anyway.

On the far side of the court were the two small buildings I had passed on my left in the early part of my tour; the open one was a Vietnamese gift shop. Then of course, the front side of the Eakin Compound Officers Open Mess (ECOOM), informally referred to as the officers club. I knew that if I just continued past the O club, went right and then left, I'd be back at the swimming pool, so I'd seen it all. I tried to peer into the club, which was not yet open, but it was dark inside, and I couldn't see a thing. Suddenly, it occurred to me that a friendly face might be a nice thing to see, and I thought of Sergeant Chavez. So I headed back to the mess hall with hope in my heart.

I crossed the parking lot and entered the breezeway, figuring to enter through the EM mess. Finding the doors locked, I continued along until I found the rear entrance to the kitchen. I was happy to see Sergeant Chavez within, especially knowing that with his presence, I wouldn't have to deal with such questions as "Who the hell are you?" or "Yes, may I help you?" or "Is there something you want here?" Self-introductions along the lines of "I'm your new boss, Lieutenant Dinan," are not my idea of the best way to begin a relationship.

Upon first stepping inside, there was a momentary surge of unintelligible conversation, or exclamation, and then audible silence. Following the general direction of the universal gaze, Sergeant Chavez came to rest his eyes upon me, smiled, and said, "I'm glad you're here. I want the pleasure of introducing you to someone."

Before I could say anything, he readdressed himself to those about him. I had no idea what he was saying, but I did pick out Dinan and Lane, and at the end of his speech, the assembled gave an outpouring of body language that clearly said "We recognize you as the boss."

I asked Sergeant Chavez, "What was that all about?"

"Basically I told them that you are *Thieu Uy*, that is Lieutenant, Dinan and that you would be replacing *Dai Uy* Lane, Captain Lane."

"Come now, you must have said something more."

"Well, I did let them know that you are much more qualified than a lieutenant usually is, and that they can be proud to work for someone as important as you were in America."

"OK, but let's not overdo this sort of thing. Now, where is this someone you want to introduce me to?"

"Follow me." I did.

We crossed the kitchen and went into the enlisted dining room. Seated at the first table to our left as we approached, was a man I judged to be about fifty years old. In his right hand was a coffee cup; in his left, a very pretty Vietnamese girl. In front of him was spread a newspaper that he was evidently reading. Inasmuch as he was wearing civilian clothing and so grandly situated in the enlisted mess at 1600 hours, my mind immediately turned to thoughts of Sergeant Blevens. I quickly concluded to myself, one Blevens per tour should be enough and silently asked, *Oh God, why me?* The man had to know that he was being approached, but he made no sign of acknowledgement until he had finished reading. It was a very short wait, but even so, at that point, I was quick to take offense at being made to wait by someone I must outrank. Before I could antagonize myself too much, the man looked up with a smile and asked, "Chavez, what do we have here?"

"This is Lieutenant Dinan. Lieutenant, I want to introduce you to Chief Warrant Officer Jay Arnold." He put down his coffee cup and offered up his hand to me in a way that said he wouldn't go this much out of his way for just anyone. I took his hand, and we exchanged a firm grip. He gripped me with his eyes as well and offered, with a very light Southern drawl, "Why don't y'all take a seat?" I did. He continued, "How'd you like a cuppa coffee, son?"

"I'd like that just fine—black," I replied.

"That's the only way a man should drink his coffee. Trong, darling," he addressed the girl upon whose hip his left hand continued to rest, "how about you get the *Thieu Uy* here a nice cuppa hot black coffee and bring a little more for your favorite round-eye?" He pushed her away and sent her off with a light smack on the backside; she rewarded him with a look that declared to all that this lady was far from fully domesticated to his will. Sergeant Chavez had disappeared, and now, with the girl gone, it was only the two of us. I was struggling to figure out what my relationship was supposed to be with this Warrant Officer sitting across from me. After all, I was a commissioned officer. Chief Warrant Officer can mean CWO2, CWO3, or CWO4. How am I supposed to know which one applies here? Do I outrank him or does he outrank me?

To clarify: commissioned officers, all of whom hold presidential commissions, always outrank enlisted personnel, including all non-commissioned officers. Warrant officers, who are soldiers with specific expertise in various military capacities, hold warrants from their Service Secretaries. However, on reaching the level of chief warrant officer, they also receive commissions from the president. Go figure!

As a young lieutenant facing an obviously highly experienced chief warrant officer with many years of military expertise, it was clear that even with my limited understanding of all military protocol in this matter, I could figure that he would have a lot more clout than I.

I remembered Sergeant Blevens saying that CWO Arnold ran the clubs and was a great guy. While I was trying to decide who should be in charge, Arnold just went ahead and took charge. "Sergeant Chavez tells me that you were a cook in New York before you joined the army. Is that right?"

"Yes." I was thinking he could have said chef instead of cook.

"That's really good to hear. There are only two members of the military who are rated as master chefs by the American Society and I'm one of them. Someone is going to be number three. Why not you some day?"

"Why not?"

That was to be the end of that subject. Rather than answer my question, the subject was changed. "General Desobry asked me to keep a particular eye on the food and beverage operations here on Eakin Compound. It's sort of a personal thing. We've known each other for a long time, and I make sure that nothing happens that might embarrass the Old Man, if you know what I mean."

I had, at this point, no idea what he might mean, but I figured, why let on? I just said, "Indeed, it wouldn't do to have the General embarrassed."

"No, it wouldn't. I've had a very good working relationship with Captain Lane that has enabled me to run the clubs here while keeping an eye on the mess and tending to my duties as Chief Warrant Officer in charge of food supplies for the Fourth Corps. What kind of experience do you have for this assignment?"

I hadn't begun to face that issue yet, and I wasn't ready to deal directly with so blunt and sobering a question. Sidestepping, I put the onus on the Powers-That-Be:

"Evidently I have whatever experience is deemed necessary by those commissioned to make these decisions." He was letting me know how well-connected and experienced he was, while shedding light on my poverty in both categories. I wasn't going to miss the opportunity to stick it to him. The *commissioned* was not lost on him.

"I see," he replied. "Say now, son, just how did you receive your commission?"

"OCS."

"I didn't realize that they had an OCS for the Quartermaster Corps. You sure you're not ROTC?"

"I'm sure," hoping that this might give me a little extra status.

"What did you do as an enlisted man—cook?"

"No. Intelligence."

"Okay then, son, let me figure that you've got plenty of intelligence, and being from New York, a little street smarts, and since you made it through OCS, you got heart. So, I'm gonna level with you. I've got twenty-eight years in this man's army and twenty-eight years has taught me that brand-new lieutenants don't know how to play ball. And when you get a player on your team who doesn't know how to play ball, the chances are you're likely to lose because that guy is going to do something stupid. I've got two more months to go here. And now I've got to watch you carefully for those two months.

"Normally, when Captain Lane leaves, you would take over his bunk, which is in my quarters, but you're not going to. If all goes well, when I leave you get my bunk. There are two quarters on the compound with air-conditioning: the General's and mine. Think about it. There's nothing personal in this. Hell, I don't even know you, but I'll help you anyway I can, if I can. It's up to you."

Just then, Trong was returning with the coffee. I was happy for the momentary diversion it provided. My thoughts on military posturing seemed really dumb to me at this point. Plain and simple was the fact that this man was connected and experienced and very probably he could help me, not help me, or hurt me, depending on his whim. I knew he was right when he said it wasn't personal; rather, it was just that no one wanted to get caught up in losing because of an inexperienced teammate. I broke the surface of my coffee with a spoon and thanked Trong while enjoying the rich aroma. I thought of what Major Hicks had said to me earlier.

"Chief Arnold, if there were some way of my starting out other than as a second lieutenant, I'd opt for it, but there isn't. If you mean what you say about your willingness to help me, I want you to know that I'm willing to accept all the help you can give me. I'm not expecting to die of it, but being a second lieutenant seems to be like an illness that I'll just have to get over."

That drew a very small chuckle. "Listen, Lieutenant Dinan. Let's take it one day at a time and see where it goes."

"Okay, Chief Arnold."

"And never mind that Chief Arnold horseshit. We warrant types don't stand on a lot of ceremony. Just call me Jay."

"Okay Jay, but only if you call me Terry."

"Terry. I gotta go now; I'll see you later." Jay was up and gone.

I finished the coffee while contemplating the fact that Captain Lane and CWO Arnold were evidently roommates who had very likely had more than one conversation bemoaning their fates due to the possible havoc I might cause. Mentally, I wrote Lane off, knowing he would be gone shortly. Jay, I wanted on my side. He came on strong, but I had the feeling that he was really one of the good guys.

NINE

AM I JEWISH?

I dropped off my cup and saucer in the kitchen and, for lack of anything better to do, returned to my quarters. As I started to unlock the door, I found the lock already opened. Inside was my roommate, stretched out on his bunk, clad in olive drab tee shirt and skivvies and sporting the war effort's answer to a golfer's tan and sun-bleached hair. He was napping comfortably until my entrance disturbed his reverie. He opened his eyes slowly and then bounded off the bed, grabbing my hand and welcoming me enthusiastically. "Hi, I'm Kirby, Kirby Goldberg from Baltimore. It really sucks here. God, I'm happy to see another lieutenant around here. All these guys are so stuck up; they treat us like shit. What did you do to receive this punishment?"

He was a real ball of energy, asking questions and giving a running commentary, leaving no room for answers. While all this was going on, I removed my fatigue jacket and sat down on my bunk. He sat on his, facing me, the monologue never letting up until something caught his eyes, stopping his mouth in the open position. He was transfixed by the chain I had around my neck. After what seemed a long time, he fairly shouted in joy, "You're Jewish! You're Jewish! Oh my God, this is too good to be true!"

He was so happy I hated to let him down, but I knew it would get messy if I didn't set things right immediately. I said, "I'm not, and it is."

"Say what? Am I *meshuggener?* Am I *meshuggener,* or is that the Star of David you're wearing around your neck?"

"It is. A good friend, a guy I went to school with in New York—his name is Rebhun, Bruce Rebhun—and he's in the jewelry business. I visited him and his Dad at their shop before shipping out, and Bruce took this off his neck and put it on mine. I've worn it ever since."

"You mean you're not Jewish?"

"That's right."

"And you don't care if people think that maybe you might be?"

"I never even thought of it, one way or the other."

"Well, if what you say is true, and I believe you, mostly because you don't look Jewish, I remain the only Jewish officer in the entire Mekong Delta, a double minority: Lieutenant and Jewish."

"Did you say you were from Baltimore?"

"Yeah."

"I was at Fort Holabird for a year or so. Not a bad city. I really liked the Towson area."

"Good, that's where my family lives. Where are you from?"

"New York."

"New York! Come on, give me a break—you've got to be a Jew."

"No. Episcopalian, and it's past 1700 hours. In accordance with my religion, I must have a drink. Why don't you put your clothes on and we'll head for the O club?" We went to the club together. I was lucky to have a guy like Kirby to show me the ropes. You know, the important things, like how to purchase a chit book (first you bought chits, with the chits you then bought drinks at twenty-five cents each), the correct bathroom to use, how to play liar's dice, and all that good stuff. Mostly, it was good to be introduced around by a peer rather than an overlord or subordinate; with a peer you could meet truly on the level.

The outstanding feature of the officers club was a very large, bright yellow poster depicting Lucy of the Charlie Brown comic strip, great with child, exclaiming, "Beaucoup 35 Charlie Brown." I tried in vain to fathom the significance of this on my own. Finally, as all those who went before me, I was forced to ask the hidden meaning therein contained. In Vietnamese, the number thirty-five is *ba-muoi-lam*, which for some reason lost in antiquity, also refers to the physical act of procreation. The code thus broken, the meaning was clear.

Kirby and I had an immediate simpatico. We didn't get to spend a great deal of time together because he had to travel all over the Delta while he was stationed at Eakin Compound, and six or seven weeks after I arrived, he was suddenly relocated, after which we lost all contact. Nonetheless, our short friendship had a lasting legacy. In the enlisted ranks there were half a dozen young Jewish boys. The highest ranking at the time was a Spec. 4. I'm not certain that any of these fellows ever made it to E-5. These guys were definitely not career-oriented, and for the most part, they felt—and in fact were—outnumbered and short of a sympathetic ear. Lieutenant Goldberg was a rabbi of sorts for these troops, and when he left, I somehow and quite suddenly inherited this position. In truth, some of the boys were convinced that I had to be Jewish and sought my advice accordingly. We would get together with some regularity, or the guys would come to me individually with one problem or another. Always there was the difficulty over food. Good—or even bad—Jewish deli simply was not available. Corned beef, pastrami, dill pickles, real rye bread, lox, bagels, cheesecake and such—these were the longed-for delicacies, and these were topics that brought great tears of homesickness to these bright, young eyes.

This is a true story, told Biblically for fun:

> And it came to pass that I went unto Hong Kong to do business. And there in that city, there dwelled at that time a restaurant with the name Lindy's, being of the same name and lineage as the famous Lindy's of New York, whence they airlifted, yea, even their cheesecake, corned beef, pastrami, bagels, lox, cream cheese, rye bread, and dills. And while I yet dwelled in that city, I did lay in such provisions as have heretofore been listed, and seeing them carefully packaged I did return laden with these gifts unto the Compound of Eakin. And in C-5, which was my dwelling place, I did assemble the children of Israel, and we did feast heartily until not one morsel was remaining. And though the law of the land, which decree-eth that an officer

shall not entertain enlisted personnel within his quarters, had been transgressed, we were not the least disturbed in our feasting. And when the hour drew nigh for the children of Israel to depart unto their own quarters, they consulted amongst themselves, and their chief spokesman reported that it was solemnly agreed that even if I were not a Jew, I was, in my person and spirit, very much one of their number.

For them and for me, it was a very special supper.

TEN

THE INVENTORY

0800 hours came around all too quickly, and as instructed, I was at the mess office attired in my jungle combat gear, sans weapon. Captain Lane assigned Sergeant Blevens the task of guiding me through a complete inventory. The army's way is very simple: when an officer signs for something, he or she becomes completely responsible for it. Should it become lost—again, very simple: the officer pays for it. Under these simple rules, one is behooved to inventory very carefully. Taking a large inventory is always tedious; at 110°F, it is unbearable. I did my best, which turned out to be pretty good, but I did discover some months later that in that other little building, the one next to the Vietnamese Gift Shop across from the volleyball court, what were supposed to be full cases of whiskey from floor to ceiling were otherwise. As it happened, this had been the last stop on my inventory tour, which had lasted all day, so I had become lazy and didn't insist that every single case be opened and inspected. Later I was to discover that the bottom two rows of cases had been emptied of booze and filled with wooden dowels. Fortunately, I was able to finesse this discovery. And so it was that I signed for thousands of dollars of goods, from Christmas tree lights and ornaments to cans of beer and, in between, all the usual paraphernalia involved with wining and dining, cooking and cleaning, and billeting and collecting.

According to regulations, all major pieces of equipment must be carefully inspected to ensure that the serial numbers match up with those documented on the original receipt. It was interesting to note that none of our major pieces of equipment had any documentation at all. It was simply amazing to me that all the stoves, refrigerators, walk-in refrigerators and freezers, coffee urns, stainless steel sinks and work tables, even the furniture in the officers mess (origin, Thailand), had simply been found on post. It was as though, when something was needed, the method of procurement was rather like a very specialized Easter egg hunt in which the interested parties merely had to search the compound with great diligence, knowing that the golden egg was always hidden somewhere in their garden. I mean, Great Finding Guys! So, for this equipment, there was no documentation. Rather, a list was made up and dated, and this list was definitively marked "Found on Post."

Day two was clearly a dud.

Day three was to be more interesting.

I spent the morning tidying up the inventories, checking figures and extensions. Not much fun, but the afternoon held the promise of adventure. Captain Lane was going to take me to Binh Thuy Air Force Base, a long and possibly dangerous trek well outside of Can Tho into the countryside. I was instructed to carry my weapon.

At 1400 hours, Captain Lane and I rendezvoused at the mess office. Both of us were packing .45's. We eased into an open-air army jeep, and we were off, buzzing out into enemy territory under arms—you know, like General Patton or something. This was really playing soldier at its best. Sitting next to the incredible hulk of Captain Lane, I had no fear.

We took a right as we cleared the gate and traveled past the MACV IV Headquarters, past the center of town with its bus station, right on past the Can Tho Army Airfield and further on into an area entirely new to me. The people, as always, were using their plumbing estuary, as I had noted before. Most just looked up and smiled and waved as we drove by; the sound of an automobile was apparently still worthy of special attention. In fact, all seemed pretty calm and normal, and Pribble had been correct: the aroma of the place was hardly a problem, even after only two days. As we continued along, we encountered fewer and fewer dwellings, and finally almost none. Open, flat-looking spaces were all around, mostly rice paddies. Occasionally, there were workers in their black pajamas and conical straw hats. Here and there, Captain Lane would point out to me areas that he judged to be particularly vulnerable to enemy attack, especially where the road twisted, narrow and pitted, with thick underbrush challenging its right to exist. I was instructed not to try this road in the rainy season; also, to check the tires and gas and to leave plenty of time to return before sundown whenever I did use it. After about an hour's traveling time, we took a left at a well-marked sign: UNITED STATES AIR FORCE—BINH THUY AIR FORCE BASE.

The Air Force did things right. The road was well paved and the brush trimmed on either side. Entering the base itself, one instantly had the sense of being on a very military installation; order prevailed. We drove by a large, gleaming hospital and pulled up in front of a substantial building adorned with Air Force crests and inscribed: HEADQUARTERS USAF BINH THUY AFB.

When we stepped inside, we effectually left Vietnam altogether and entered into a beautifully air-conditioned American Air Force world. Nothing about this place would have been out of place in Omaha, Nebraska, or anywhere else in the States for that matter; that is, with the exception of Captain Lane and myself who were undoubtedly the sweatiest, road-dustiest things to enter this pristine environment since Captain Lane's last visit.

Nonetheless, as ever, Lane was well recognized and well received. We made our way to Captain Northrup's office where Captain Lane received an adventurer's welcome: "Look at you, you made it through again!"

"No problem."

"Well, you wouldn't catch me on that road outside of an APC (Armored Personnel Carrier)."

"Come on, let me take you for a spin. It will do you good."

"No thanks. You ground troops can keep the ground around here for yourselves. I'll stick to the air."

"OK, have it your way, but this might be your last chance. I'm shipping out. This here little Lieutenant is taking over my desk."

Captain Northrup was no poker player! How can a face so clearly shout

"You must be kidding" that loudly and yet not make a sound? This naked reaction scored a direct hit on Captain Lane as well, and to his credit, it made him uncomfortable; he did try to set things up for me.

"This is Lieutenant Dinan. Dinan, I want you to meet Captain Northrup."

I answered, "A pleasure to meet you, Sir."

"Ah-um-ah-oh-um-ah-yes, thank you, yes," he replied. Clearly this one was suffering from culture shock.

Captain Lane went on, "I've taken the liberty of explaining to Lieutenant Dinan the relationship I've had with you and your predecessor. I've explained the type of help you have been to me and . . ."

Captain Northrup interrupted, "I certainly hope you made it clear that I'm not required to give any support to MACV."

"I said it was a favor, and by the way, one of the reasons I've made this dangerous drive all the way out here is because I wanted to thank you face to face for your many kindnesses. You sure have saved me a lot of trips to Saigon."

"Oh come on, Ed. You know I've been happy to help you."

"Thanks, Bill, and I hope you'll be able to bail out my replacement here if he gets stuck."

The direct request was like a fist to his solar plexus. Air actually escaped his lips, and his eyes bulged. I couldn't figure what the big deal really was. As Captain Lane had explained it, one of my duties was to always have plenty of Vietnamese piasters on hand to exchange for military payment certificates. Since the GIs were not allowed to spend this military scrip on the local economy, the Eakin Compound piaster exchange facility was a very important link in the chain of that regulation. Of all the possible sins of omission that I could possibly commit, running out of these piasters was far and away the most unforgivable. It was so unthinkable, in fact, that the appropriate disciplinary action in the event of a failure to comply had not been either conceived or communicated. That some things "shall be done!" is simply taken for granted, and this was such a thing.

The money exchange was a one-way affair. I was to facilitate the exchange of MPC for piasters but never, *never,* piasters for MPC. The supply of piasters could only be replenished at authorized resupply facilities, and the major facility was the bank in Saigon. In an emergency, I could use another exchange facility such as the one at Binh Thuy. Since the trip to Saigon was a miserable and, at minimum, day-long procedure, this trip into the countryside could become as habit-forming for me as it had for Captain Lane.

Captain Northrup was coming up for air. "Well, really, Ed, you know, between us it was one thing—and of course, in a real emergency—well I, naturally, would always do my level best to help another member of the American Armed Forces. But I can't promise to be available to every second lieutenant that lets himself get behind the eight ball."

I wanted to tell him to go and fuck himself, but I figured I might really need his help one day, so instead I said, "Gee, thanks, Sir. You know I'll try to never get stuck behind that eight ball, but it's a real relief to know that should I stumble, there's a helping hand here to keep me from falling."

With great condescension he added, "Once, maybe twice, but don't get to rely on it—and don't show up without calling."

This was becoming far too much fun. An exit line was in order. "Yes, Sir. Thank you, Sir. Could you direct me to the men's room?"

Captain Lane interjected and took me to the door, gave me directions, and asked that I meet him by the jeep.

I did the men's room and then made my way back to the jeep. After being spoiled by the Air Force's wonderful air-conditioning, the heat seemed doubly depressing. With not a stir in the air and the sun blistering down, it was *hot.* Happily, Captain Lane came along quickly, and we were off making our own breezes. Captain Lane seemed a little glum, but I didn't realize that we hadn't exchanged so much as a word until he began talking: "I'm sorry that you didn't get a better reception from my buddy back there, but don't take it personally. You see, virtually no one is very interested in doing favors for those they outrank. In this man's army, you kiss up, not down."

"I believe I'm getting that message, loud and clear. Do you suppose I should write up a recommendation that at least a couple of hours be spent on teaching this reality in all officer training courses?"

"My recommendation, Lieutenant, is that you be very careful about anything you put in writing. Get *that* message loud and clear!"

"Yes, Sir."

The trip back was uneventful and rather pleasant. The foliage was lush, and the feeling of being in a different, far away, remote place still colored each new observation with particular excitement. When we pulled into a parking space in Eakin Compound and came to a stop, Captain Lane turned to me and said, "Well, Lieutenant, it's almost 1700 hours. How about you put up your weapon, get cleaned up, and meet me in the officers club? I think it's time I bought you a drink."

I was thrilled. This was a major breakthrough. "Yes, Sir. Thank you, Sir. I'll be waiting for you."

I washed up and made my way to the club, hoping that with the aid of a couple of relaxing cocktails the Captain and I might approach each other on the level. I checked inside and found that Captain Lane had not yet arrived. I went back out to wait for him so that we could enter as a team.

Shortly, he approached with Jay Arnold at his side. "What are you doing out here?"

"I thought I'd . . ."

"Never mind. Let's go in so that I can buy you that drink."

We went in and took a table. A little Vietnamese waitress came over, blushing with a shy smile and little bows; after all, at this table was the power of the Can Tho Mess Association. Captain Lane ordered a vodka and tonic, so I did likewise, and Jay ordered a mai tai with rum. While the drinks were coming, Captain Lane, with great seriousness, told me, "By the way, if I were you, I wouldn't depend too much on the exchange facility at Binh Thuy. Things are changing around here, and there are a lot more flights to Saigon than there used to be. So you should be okay."

"Thank you, Sir. I won't."

When the drinks arrived, Captain Lane raised his glass and explained,

"It's a tradition here that everyone buys a drink for his replacement. I am happy to be continuing this tradition." We all drank to that, and our glasses were quickly emptied.

"And now," I offered, "let me have the honor of buying you gentlemen a drink."

"Sorry, but I've got a letter to write," Captain Lane declined. "I'll see you in the morning." And with that he was gone.

Jay, seeing that I was miffed, said "Hey, Terry! Forget it, kid. This guy writes a letter to his wife every single night. I guarantee he hasn't missed a day."

My thought was: *Bully for him. I guess he can tell his wife what a good boy he was fulfilling tradition in the face of such hardship.* So much for the opportunity to meet on the level.

ELEVEN

MR. TRAN VON HEIN

ON my fourth day at Eakin Compound, I was engaged chiefly in the business of signing my life away. With the application of my distinctive signature to varied and voluminous official documents, I at once became personally responsible for tens of thousands of dollars of inventory, equipment, and actual money. Since the cumulative sum of these assets far exceeded my expected earnings for the foreseeable future, this was a disquieting experience.

My day was further disquieted when I noticed that an inordinate number of my fellow dwellers sported frightening-looking plaster casts on various limbs accoutered by crutches, slings, and whatnot. Although I had been informed that the mortars and rockets that arrived unwelcome into our little space had caused no deaths to date, the prospect of a crippling encounter was not to my liking.

I was later to discover that these were not war wounds but rather sport wounds. More specifically, they were the result of volleyball. Sunday was Volleyball Day. Rain or shine, high-intensity competitive volleyball was in order—and in the Delta, rain was a rather common phenomenon. Often, the local Vietnamese team was the competition. Our athletic director had been a runner-up for a spot on the U.S. Olympic Team, and several of my compatriots had represented West Point or some other distinguished place of higher learning. Played with what seemed to me to be reckless abandon on a concrete surface, these matches would, under ideal circumstances, have justly caused many injuries. In the rain, however, unjust injuries were common. To add a little spice to the activity, it was the custom that the losing side would sport the winning team to refreshing adult beverages between sets. Here we have achieved a script for disaster. Combine wetted-down players with a wetted-down concrete playing court, let them compete with noble abandon, and you will have broken bones—guaranteed.

The high point for the day was my introduction to Mr. Tran Von Hein. Along with acquiring responsibility for a lot of stuff and functions, I acquired responsibility for a staff comprised of 15 U.S. Army enlisted personnel and 150 Vietnamese civilians. The most important civilian on my payroll was, without question, Mr. Hein.

Captain Lane made the introduction: "This is Hein," he said. "Hein is my liaison to the Vietnamese staff. You will find him very helpful. Hein, this is Lieutenant Dinan. He is my replacement."

It was altogether a rather perfunctory introduction. We shook hands, and I was welcomed with assurances, in perfect English, that he would do his best to be helpful. As I had many duties to fulfill attendant to my replacement activities, I did not have time to engage my liaison-to-be in lengthy conversation at that time, but I did notice that the Vietnamese staff treated him with great deference. The U.S. staff, on the other hand, deported themselves as his superior. At the time, it was not a judgment, but rather a strong impression. I did not dwell upon this, as I had more pressing activities to occupy my attention, but I did take notice. As I became more acquainted with Hein, he became for me, and under my direction to all my staff, "Mr. Hein."

Although Mr. Hein possessed none of the physical attributes associated with aristocracy, there were about him the telltale signs of a person who has gained personal respect by dint of personal achievement. In Mr. Hein I saw that air of self-assurance, the posture, the poise, the indefinable something that instantly communicates to the attentive observer that he is in the presence of someone who is worthy of respect. Tran Von Hein had been an anti-Communist reporter in the north of Vietnam. As a Roman Catholic with sufficient power of the pen to inspire a reliable following, he became a prime target for assassination following the Communist takeover in the North. With no more than the clothes on his back and a considerable price on his head, he found his way as far south as Can Tho in order to avoid certain death. Arriving with nothing save his intellect and abilities, he quickly achieved prominence in this important provincial capital city.

By the time I arrived in Can Tho, Mr. Hein was the proud father of nine children and the sole means of support for two households that included his in-laws and a domestic staff of eleven. Included in the eleven was his personal chauffeur who was placed at my disposal by Mr. Hein on numerous occasions. While I was in-country, Mr. Hein purchased a new vehicle from Sweden. It should be noted that on top of the usual purchase price, the imported vehicles carried a 200 percent import duty. Ergo his US$10,000 car cost him $30,000 cash, and this was in an economy where the going minimum wage was set at $0.25 per hour. It should also be noted that his purchase was completed just in time to avoid paying the new import duty of 300 percent instituted by the Vietnamese president. This directive or legislation also stipulated that the import duty on the purchase of a certain make of Japanese motor scooter was set at zero. I was told that for each scooter sold in Vietnam, the president received a bounty of US$15, and further, that more than one million of these scooters were purchased in 1968 alone.

Yes, by any standards, Mr. Hein was rich and successful. Along with being completely literate in French, English, and Vietnamese, Mr. Hein held the distinction of owning and operating the only bakery and the only fresh vegetable processing plant in Vietnam that were fully approved by the U.S. inspectors. He also owned a laundry service and God only knows what else. The General's staff at MACV IV Headquarters pressured him continually to sign on as their chief translator, but he wisely refused, as it might have caused allegations of gross conflict of interest. When an important translation was needed, Mr. Hein would perform the task but refuse any payment. As a consequence of the military establishment's inability to gain control over Mr. Hein, he was both revered and reviled by the MACV IV Headquarters staff. In either event, his skills were necessarily engaged as necessity dictated.

After I got to know him better, I told him I felt I should be working for him, rather than he for me. He allowed that I was mistaken, but he understood and appreciated the fact that I held him in high regard and respected him as an executive. When he tried to leave my employ because he feared conflict of interest allegations, I was able to keep him as an adviser on a retainer. In agreeing to this arrangement, he stipulated that when I went, he would go also. I assumed that would be a negative for my replacement, but by that time it was a very different job. Actually, when I was replaced, the operation was broken up, and I was replaced by four captains. The point is that Mr. Hein was a phenomenon, without whose able assistance my job would have been much more difficult, if not impossible.

Can Tho Army Airfield

Can Tho Army Airfield

Living on the Bassac River

Children playing in Can Tho

Can Tho traffic: hand-drawn cart, bicycle, U.S. Army truck

Eakin Compound parking area

Private Pribble relaxing by the Can Tho Mess Association Office (sign obscured at right) and the basketball half court

Kitchen staff waiting to pick up a la carte orders

Officers mess/dining room

PX, left, and Row C, right

Lieutenant Dinan with a pool party set-up

Lieutenant Dinan at his desk with sign: BLESS THIS MESS.

Private Pribble in the cashier's cage with sign:
PERSONNEL IN CIVILIAN CLOTHES MUST SHOW ID CARD.

Cooling off in the pool despite the barbed wire atop Eakin Compound's outer wall

Lieutenant King and Lieutenant Dinan after badminton

Lieutenant Dinan enjoying the basketball/volleyball court

Sunday volleyball—on a soaking wet court as usual

Lieutenant Dinan indulging
in the sun and the pool

A rugged game of
four-wall handball

Row C's covered walkway and sandbag bunker

TWELVE

A "REAL" WARRANT OFFICER

CAPTAIN Lane was soon on his way, and I was in charge—of what? The Can Tho Mess Association consisted of the following: the general officers lounge, the officers club, the senior NCO club, the enlisted mens club, the officers mess (which included the generals table), the enlisted mess, a snack bar, the piaster exchange facility, and billeting services for Eakin Compound (rather like hotel services: the making of beds, the cleaning of rooms and facilities, and all things of that sort). Also at that time, we sold food to five teams out in the field. To provide for all the attendant needs, I had, as previously mentioned, a staff consisting of 15 U.S. Army personnel and 150 Vietnamese nationals.

Chief Warrant Officer Arnold had been running the clubs, and he was also intimately involved with the operation of the dining rooms. As it happened, these were not rightfully his responsibilities. His job was to be responsible for all food-related issues in the Mekong Delta, of which Eakin Compound was only a very small part. For whatever reason (he had known General Desobry for a long time, and Open Mess Management was his Military Occupational Specialty (MOS)), he had spent most of his time doing much of Captain Lane's job for him. This arrangement quickly ended following my takeover, and CWO Arnold spent his last two months in-country looking after the General and the Delta feeding functions. I became the sole commissioned officer looking after the Can Tho Mess Association.

It was a big responsibility and really a great deal of work. Oddly enough, my first job performance rating as commissioned officer was performed by a lieutenant colonel. This is not supposed to happen to a second lieutenant. As it turned out, Lieutenant Colonel Bagley was the G4 of the Fourth Corps. That is, he was the Quartermaster Officer in charge of all QM activities in IV Corps and a member of the Command and General Staff. Because of his habit of keeping a diary of all noteworthy events at meetings, he was much feared by his peers. They often found a true recording of events inconvenient to their wish to creatively reconstruct the nature of their actual participation. When it happened that he actually had to submit a rating for me, he explained that he had taken three points off a perfect score because, "A second lieutenant should never receive a perfect rating, and besides, I clearly recall mentioning the need for a haircut to you on three occasions."

One of those occasions is subject to a high level of recall. One afternoon, as I was cutting across the parking lot for some important purpose, who should be standing at its very center but Colonel Bagley, looking very smart in that military way, standing erect in heavily starched fatigues. In his left hand he held his ever-present diary. In accordance with custom, the right hand was kept free for possible saluting purposes. The diary, a very handsome leather-bound tome comprised of permanently attached pages, was his evident weapon of choice, and he was rarely seen unarmed. While engaged in scanning various horizons like a military observer, he took notice of me and summoned me into his immediate presence.

As I looked into his crystal blue eyes, I experienced the feeling that he was looking more into me, rather than at or through me. This particular ability, seemingly restricted to the best of senior military officers, causes one to experience the feeling of being unclothed. Its affect is unhinging without being mean-spirited.

Thus paralyzed, I was addressed: "Lieutenant Dinan, you should be interested to know that I am required to submit your OER (officer's efficiency report), and it is highly unusual for someone of my rank to rate a second lieutenant. Actually, the last time I did such a thing, I was a captain, and that was a long time ago. Nonetheless, it is required and in this case I am happy to do so. I was therefore thinking about you and your assignment. As I stood here looking about, I noted as I looked that Lieutenant Dinan is responsible for this and Lieutenant Dinan is responsible for that, and that, and that, and that—and, from an organizational standpoint, a lieutenant should not be allowed to be responsible for all that much. I am not saying that you are not doing a good job of it. As a matter of fact, I find that you are doing an excellent job; it's just that the job you are doing should be done by a major or a senior captain with a minimum of two captains or senior first lieutenants to support his efforts. Do you understand what I am saying, Lieutenant Dinan?"

"I think so, Sir."

"Listen, if you should ever find yourself overwhelmed, come to me and I will find a way to help you."

"Yes, Sir."

"Do you understand what I am saying, Lieutenant Dinan?"

"I think so, Sir, and thank you, Sir."

"Very well, and, Lieutenant Dinan, you need a haircut."

"Yes, Sir." I had my hair cut that very afternoon.

Approximately six weeks later I made an appointment to see Colonel Bagley at his office in the MACV IV Command Headquarters. This was a distinctly separate facility from Eakin Compound, which was dedicated to human wants and needs like sleeping, eating, playing, and a host of similar frivolous pursuits. Located in downtown Can Tho, this was MACV IV, Military Assistance Command Vietnam Fourth Corps. This was where the serious game of war was played. Well, actually, the serious game of advising on how war should be played, since IV Corps, the Delta area, was designated as the military playing field for the Vietnamese Armed Forces. Alas, we were not here with the primary mission of doing any actual killing; rather, we were here to instruct others on how they might best accomplish this important activity. A second lieutenant is assumed to know nothing and to be incapable of giving sound advice. Therefore, it is understandable that on these sacred grounds the foot treads of such unworthy souls were not to be heard. Here everything was rank, rank, RANK. The enlisted component was plush with up and down stripes, including a no less notable personage than the Sergeant Major of IV Corps himself. The officers were all field grade and above (majors, colonels, and generals), with the exception of a few captains occupying the positions of aide-de-camp or some other supporting role.

This was my first visit to the MACV IV Command Compound, and I could not help but be impressed. Arranged in the shape of a *U* around a central parade ground-cum-parking field, a series of imposing entryways were uniformly identified as to their military function using standard military nomenclature: MACV IV G1, MACV IV G2, MACV IV G3, MACV IV G4, MACV IV IG, and of course, MACV IV HQ—aka Personnel, Military Intelligence, Operations and Training, Logistics, Inspector General, and, of course, the Commanding General.

I headed for and entered under the entryway advertised to be MACV IV G4 where I was received graciously, if somewhat curiously, by the quartermaster personnel assembled therein. Having duly noted my golden bars, second lieutenant insignia, they summarily assessed my worth and chose to ignore me. Interrupting whatever ponderous, monumental task he was in the process of pretending to be engaged in, I introduced myself to the staffer, a sergeant, closest to the entry: "I am Lieutenant Dinan. I have an appointment to see Colonel Bagley."

Looking up at me from his desk with an expression that screamed both annoyance and defeat, he capitulated to his duty and instructed me, "Lieutenant Dinan, you will take a seat and I will inform you when the Colonel will see you."

Avoiding further contact, he immediately turned away and headed to a door in the rear of the room. With unexpected swiftness, the sergeant returned and instructed me, "Lieutenant Dinan, you will go in now."

As I approached the door, I saw it was open, and I could see that Colonel Bagley was standing behind his desk. I didn't know if I should report in the prescribed military manner or what, but since the Colonel instantly took charge by welcoming me and extending his hand, I recognized that those formalities were deemed unnecessary.

I took his hand rather than saluting and said, "Thank you for seeing me, Sir."

"My pleasure, Lieutenant Dinan. Have a seat and tell me what's on your mind."

I sat myself quickly, recognizing that the offer was more of an order than a gesture of hospitality. We seated ourselves in unison (who's timing?). In the process I noticed that his desk was devoid of all objects save the *book,* discreetly positioned just right of center, with a gold Mark Cross pen placed at an exact forty-five degree angle across its fine leather binding.

As we faced each other, his expectant countenance indicated forcefully that I would have to begin the dialogue. And so I did. "Colonel Bagley, I have been thinking about what you said to me several weeks ago, and I've come to the conclusion that what you said is absolutely correct." Here I paused, waiting for him to say something that would help me to further pursue this topic in somewhat more detail. He was not accommodating. As he continued to look at me benignly with a little glint in his eyes, I understood that the ball was still firmly in my court.

So I made my case. "Now that Jay Arnold's relinquishment of operational involvement in the mess association has had time to make clear the extent of my duties, I find that, as you noted, the job is really too much for one officer. If there is some way that you could find a qualified officer, one who I would actually outrank, who might assist me in this operation, I would be very grateful."

I had assumed that the Colonel would respond by asking me something specific, anything. He did not. Instead he simply said, "Thank you for bringing this to my attention. I will see to the situation." Interview over.

I left knowing that action would be taken. Eventually it was, but the results proved to be very much other than I hoped for or expected.

I returned to my office where Sergeant Sullivan was strenuously engaged in completing the end of the month financial statement. Here was an E-7 who had recently been refused a promotion to E-8 because he was sidelined to the mess association instead of working in his MOS, which happened to be postal services. Faced with a maze-like financial document set to annotate, he was clearly highly agitated.

I made an attempt to offer assistance, but I was brushed off with the admonition, "Lieutenant Dinan, I know that you mean well, but this is an enlisted man's work. Captain Lane never got involved in this, and I don't think you should inject yourself into it, either."

I said, "OK" and let it go for the time being, but I was curious.

Ultimately I was presented with the finished product consisting of page after page of incomprehensible numerical recordings that I, with some misgivings, dutifully approved with my signature. Following the submission of this document, I spent a considerable amount of time trying to make sense of it. Although the document ultimately showed an activity's profit or loss, it provided little or no information pinpointing the reasons for either outcome. This was evidently planned by the originators of this report, and as such, it was an exercise in futility from an operational standpoint. To explain: when we fill out our income tax forms, the format allows an accountant to compare certain numbers to a prevailing standard that enables the accountant to note inconsistencies immediately, thereby alerting him to the necessity of further inspection. Although more than adequate from a preventative vantage point regarding theft and stupidity, this army recordkeeping system provided no desirable operational benefit. I suspect that this was purposeful, but who knows? Right then and there, I became determined to have a system that would acquaint me with the practical reality of what was actually going on.

With this goal in mind, I labored greatly to convert our recordkeeping system into the standard double entry accounting system used in the hospitality industry. It may seem an exceptional ability to those unfamiliar with accounting, but once my books were organized I could monitor every aspect of my operations from the position, if not the comfort, of my desk.

"Real" warrant officers, that is to say those relatively rare individuals who have in their enlisted capacity so distinguished themselves in their chosen

MOS that they warrant special recognition, are afforded a very special place in the military hierarchy. In no way should those who have achieved warrant officer rank through special training, such as flying helicopters or other highly specialized technical activities, take offense, but they are *other* and easily recognized as such by a usual age differential of twenty or so years. CWO Jay Arnold was a "real" warrant officer.

The reality is that this particular group of military operatives had acquired such an expertise in the arcane world of forms, formatting, and the manifold peculiarities inherent to the successful submission of documentation related to their various areas of influence, that they were not only held in high esteem, but were in equal measure feared by the commissioned officers whose activities placed them under a warrant officer's operational influence. The enlisted ranks, on the other hand, confined their attitude to respect only. The result was that whatever a warrant officer was able to accomplish in terms of personal comfort was considered to be a perk of position beyond questioning. For example, the air-conditioned singularity of Quarters C-5, where I was also quartered for most of my tour, was never challenged *because* a warrant officer was always in residence. When CWO Jay Arnold shipped out, he was replaced by CWO "Chief" Graham.

Although I had instant respect for CWO Arnold, and though he was always very cordial to me, I sensed that his cordiality was more military than personal. I had attempted several ways of interacting with him that I'd hoped would result in some signs of personal or real respect, but nothing seemed to work. I felt like I was looked upon as just an inexperienced kid who had to be tolerated by this highly experienced and respected adult.

Late one afternoon I settled upon a new tactic. I approached him and asked, "Jay, do you have any particular plans for this evening?"

"Nothing very special. What's on your mind, Terry?"

"I know that you'll be shipping out very soon, so I don't have much time left to get your advice about running this show. I'm hoping that you could sit with me privately for a couple of hours so that I can get some real advice from you. My roommate Kirby is away, so maybe my room would be convenient. I promise to have some fine cognac on hand."

"OK, I'll grant you this one opportunity. I'll be there at 2030 hours, and the cognac had best be fine, indeed."

I thanked him and went directly to the PX to purchase a bottle of Courvoisier VSOP. Then I went to my quarters and tried to make the place appear hospitable. At 2030 hours I was standing at my opened door awaiting his arrival. Naturally, he was on time and following brief greetings, we went inside and closed the door. I had set a footlocker in place between the two bunks and arranged the brandy and glasses on a tray. Seated on opposite bunks we began talking and sipping the cognac, which had met with his approval, although he did mention that he really preferred Martell Cordon Bleu. Had I only known! I was consistently more generous in adding to his glass than to my own. Soon things became more relaxed, and the conversation became more informal, friendly, and fun. By the time the bottle was empty, three hours had passed, and we were chatting it up like long-lost buddies. Jay noted that an empty bottle was a signal that it was time to go, and off he went on less-than-steady legs. Just after the door closed, I heard the distinct sounds of garbage cans being greatly disturbed, and as I peered out surreptitiously, I could see Jay was on the losing side of this engagement. I further noted that he was expressing his dismay in the most colorful of language. The subject of his after-meeting battle was never broached, but he did mention that I may possibly have been over generous on his behalf. I stuck to the story of a fifty-fifty split on the brandy and the subject was let go. Happily, I can report that the tactic proved to be an unqualified success, and during his remaining time CWO Arnold was indeed very helpful to me.

Jay's MOS was Open Mess Management, which in practice is operating military clubs. His considerable talents in putting together lunches, dinners, and assorted functions were very much honored and appreciated by the General and his staff. As the MACV IV Command Compound was the Headquarters for the entire Fourth Corps and everyone assigned to the Command Compound was billeted on Eakin Compound, high-level entertaining was usual and required for visiting dignitaries of all nationalities, both civilian as well as military. Jay tipped me off to the fact that there was real concern at the highest levels regarding my ability to perform adequately in this important area. To allay these concerns, I proposed to Jay that I should take charge of a grand dinner honoring his service and departure, and that his responsibility would be to provide the guest list and to ensure attendance. He thought this proposal was the very best of ideas, and so it came to pass.

THIRTEEN

THE GRAND DINNER

THE first order of business was to determine exactly how many people the officers dining room could best accommodate at one grand table. I stayed in the dining room the next evening until all the diners had finished up and left. It was rather late, since the General and his staff were at that time in the habit of arriving following lengthy pre-dinner libations in the general officers lounge. Following their eventual departure, and with the assistance of several of the enlisted staff, I moved tables and chairs about until I was satisfied that the very best arrangement had been made. The magic number turned out to be eighteen, with three at the head, three at the foot, and six along each side. I turned this information over to Jay the following day, reminding him at the same time that he was in charge of the guest list and the seating. He inquired regarding the menu, and I told him truthfully that I had no idea. That decision would depend upon what I found available when I went to the main PX in Saigon. Later, after doing my shopping, I decided to keep the menu secret in order to achieve the greater impact.

It was my first trip to Saigon for purposes of bank deposits, piaster exchange, and checking on our supplies held at Tan Son Nhut Airfield. Of course I was apprehensive, but as I was accompanied by Sergeant Blevens, I had an excellent and experienced guide, and therefore everything was well organized. These trips were anything but easy. First, one had to get to Can Tho Airfield and then find a flight going to Saigon. Once at Tan Son

Nhut—with its ever comforting signage: IN CASE OF MORTAR ATTACK DON'T PANIC DON'T RUN—LAY DOWN ON THE FLOOR AND COVER YOUR HEAD WITH YOUR HANDS—it was necessary to secure transportation to the Cholon section, where the banks, exchanges, and PX were located. After completing one's business, the reverse process would begin. Cholon to Tan Son Nhut to Can Tho to Eakin Compound, all the while carrying heavy money bags and burdened with weapons required to protect same and self.

To merely say I was impressed by the plethora of food stuffs available would be a serious understatement. The quality and the quantity were astonishing. I was flabbergasted. The grocery stores I had used back home in New York were in every way dwarfed. Everything, literally everything, was available for purchase. The COLA operatives in this city were well cared for. At length I selected two sirloin strips and four sides of Pacific salmon for our upcoming extravaganza.

We had a machine for making sherbet in our kitchen, but I was unhappy with the finished product, so I had written a letter to my civilian mentor, Jerry Berns, at The "21" Club in New York, requesting a copy of the "21" recipes. In answer, I was presented by the mail clerk with an elegant airmail envelope embossed with the Marine Corps emblem and announcing the sender to be Colonel I. Robert Kriendler, USMC. A short note from Mr. Jerry advised me that he had turned my request over to Mr. Bob inasmuch as this was a military-related request. Mr. Bob, aka Colonel Bob, was also an owner of "21" and the president of the corporation. Lieutenants do not as a rule receive personal mail from colonels, and over time our correspondence caused not a few eyebrows to rise, which was kind of fun. In this first letter, I received the recipes and they were excellent.

Armed with the knowledge that the local vegetables were excellent, varied, plentiful, and under Mr. Hein's control—and therefore fully available to me on a "when I come, I'm first served" basis—I was prepared to finalize the menu. I determined that the opener, or first course, would be *Poached Salmon Mount Vernon.* Should you be unaware, this dish, believed to have originated in the Virginia home of General George Washington, consists of salmon poached in white wine and served cold, accompanied by a pressed cucumber salad lightly dressed with a sour cream dressing, sauce verte (a green sauce that is mayonnaise-based and greened with a purée of watercress, spinach, parsley, and other available green condiments), and slices of fresh tomato. This was to be served with a fully chilled white wine.

The next course, or intermezzo, was a lemon sherbet, "21" Club recipe, served on a bed of minced local pineapple and topped with likewise indigenous strawberries. This would be served with fully chilled champagne.

The meat course was *Roast Sirloin of Beef Bollinger avec les Asperges.* The roast sirloin part is almost self-explanatory. The entire sirloin strip (that's the part from which you get New York cut sirloin steaks) is simply seasoned with salt and pepper, roasted until the center is rendered the most appealing shade of pink, and then cut into three-quarter inch slices for serving. *Bollinger* refers to the accompanying potatoes that are peeled, thinly sliced, and roasted along with thinly sliced onions in a beef broth. *Avec les Asperges* means with asparagus that are fully peeled in the French style and cooked until just done. These would be served on a separate plate. All of this would be enhanced by the pouring of the finest red wine we could muster.

For dessert, I chose *Baked Alaska Flambé* to be served with a chocolate sauce followed by coffee, cognac, and cigars. Altogether, it was something of an upscale menu and most probably the very first Baked Alaska ever to be served in the Mekong Delta.

On the appointed evening, we labored greatly to set the great table perfectly. We had to improvise tablecloths by using bed linens, but we used new ones, and they were pure white and crisp. The five glasses at each place setting for water, white wine, champagne, red wine, and brandy were set using the string-guide method. That is where you draw a string from one end of the table to the other, thereby ensuring a straight line for the alignment of stemware. Lacking sterling, we were forced to suffer along with stainless steel flatware, which numbered ten pieces per place setting. This, too, was perfectly aligned. Other adornments included place cards, set in accordance with CWO Arnold's instructions, an abundance of candles, and centerpieces produced by Can Tho's finest florist. A flag bearing one star was placed beside the General's place card.

The dining room looked absolutely fabulous, inviting, and enviable. At the appointed hour, 2100 hours, the guests, having been pre-greased in the generals lounge, arrived en mass. General Desobry, all decked out in combat fatigues (it really was a dress blues occasion, but such is war) and leading the entourage, immediately discovered his flagged place card at the center seat at the head of the table.

Upon observing the place card to his right, the General stood and announced a change in the seating. "I note that Colonel Davis is to be seated at my right, which, although in correct accord with military custom, is wrong for this evening. I want Chief Warrant Officer Jay Arnold to occupy this seat of honor at my right."

This was in no way a suggestion; it was an instruction, and that which was instructed was carried out. Thus Jay was seated directly to the right of the General. This was to prove to be a somewhat indelicate arrangement before the meal ended.

All went along smashingly. The manifest pleasure of each participant was, in every sense of that word, evident. The exchange between courses was flawless, serving from the left and clearing from the right. The wine flowed seamlessly, with me acting as head sommelier. When the meat course was cleared and all the condiments needed for the main meal—salt, pepper, butter, and the like—had been removed, the dessert service was positioned for use. The time to serve the Baked Alaska was at hand.

Creating this item was a particular challenge because we were unable to find on post an electric beater to assist us in preparing the meringue, and the consequent necessity of hand-beating this egg white and sugar concoction exhausted several of our staff. Notwithstanding, a perfect meringue for the purpose of a Baked Alaska was accomplished. Meanwhile, I constructed the inner core: ice cream surrounded by thinly sliced plain white cake. The precise shape of the finished product is highly variable and subject to the discretion of the various interested and involved parties. Having assumed full responsibility for this discretion, I formed the product in accordance with the particular shape used for bachelor parties at my former place of employ, "21." That particular shape is the male member in full bloom accompanied by the usual equipage. The size of this representation when intended for the consumption of eighteen men can be easily imagined. Following the application of the meringue and the detailed sculpting of various particulars, I then used the remaining meringue to script over the Baked Alaska's extraordinary length, "Good Luck Jay."

Now was the moment: the Alaska, toasted to a glorious golden brown, was set afire with warmed brandy. This glorious spectacle, emitting all the wondrous scents and visual sumptuousness available to its flaming voluptuousness, was paraded in and placed directly in front of the guest of honor. As noted, to the left thereof was the unplanned positioning of the Commanding General.

Upon observation, the General first asked, "What is *that*?"

CWO Arnold quickly responded, "Straight arrow, Sir. It's a straight arrow"—which in the military is a common reference to a soldier who has accomplished his tour of duty with sexual innocence.

General Desobry looked at this masterpiece once again and with notable agitation declared: "I know what that is! Get it away from me!"

The monster, still lightly flaming, was removed and then served. I noted that the General enjoyed the eating of this at a level equal to his discomfort in looking upon it. Following the dessert, coffee, cigars, and Martell Cordon Bleu cognac (what else for CWO Arnold?) were offered and served. A grand time was had by all, and in spite of the little discord over the shape of the dessert, the entire senior staff understood that I was able to attend to their most ambitious entertaining needs.

FOURTEEN

COON SKIN DAVIS

THE next person of great importance to ship out was Colonel "Coon Skin" Davis. Coon Skin—my word, the Colonel did relish that cognomen. He was in every way a Son of the South and in every way he gloried in that blessing. A big man who sported thinning red hair, the Colonel carefully oiled and combed the few remaining topmost strands from left to right over a large, well-shaped, pink, and otherwise hairless dome. Somewhat overweight in the style of a professional boxer or wrestler no longer in training, he appeared, while standing still, to resemble a bear. When he moved, however, he displayed the grace and assured movements of a cat.

The first time I encountered Colonel Davis was at the bar in the officers club. As he entered, there was an automatic transfer of all attention to his person.

"Boy!" he exclaimed, as he suddenly encased a lieutenant colonel in his entourage in a one-armed bear hug, "I'll have a MAI tai!"

As the vibrations caused by this exclamation, delivered in a bass-baritone voice enriched by a decidedly southern drawl, enveloped everything and everyone, the Colonel commandeered a table, and the lieutenant colonel dutifully went to the bar. There is unquestionably a certain joy in being present when an officer of rank is addressed as *boy.* Naturally I was dying to learn just exactly who this was.

Colonel Davis was the Deputy Commander, MACV IV. As such, he was the number two military man in the Delta. He was also a legend, a soldier's soldier, loved, respected, and feared. A leader from the old school, he never was known to demand more of others than he gave of himself. You just knew that he knew that he could outrun and outgun each and every soldier under his command. This assessment was universally accepted, and rather than inspiring any sort of jealousy, it inspired that pride of association in the troops that is unable to even enter into the dreams of merely good leaders.

By the way, a mai tai is a fruit drink concoction heavily based on the pineapple, which was popularized in Hawaii. To be served properly, it must be iced and liberally infused with the finest of rums. How or why this became Coon Skin's favored libation is unknown, but it has been speculated that the very curling of those words in his mouth provided such pleasure that the ensuing joys were but secondary. I was formally advised by CWO Arnold that of all the items we might suffer to run out of, the running out of mai tai mix would be insufferable.

Colonel Davis, who evidenced a low regard for a good night's sleep, loved good military companionship and a good card game. In order to accommodate the Colonel's particular disregard for tucking-in at some reasonable hour and at the same time observe regulation hours for closing the O club (0100 hours), a special card room was formatted and fitted with an outer and inner door. The inner door opened into the club and was lockable from the club side. The outer door leading into the compound would self-lock upon closing. Between these two portals there was a specially constructed octagonal poker table, complete with felt topping and the customary wells for chips and refreshments. This was the Colonel's domain.

One part of the Coon Skin legend incorporates this domain. Late one night, or more accurately, early one morning at about 0230 hours, the officer of the day (that's the one who has to stay up all night) received an emergency call regarding a sub-sector team in dire straits due to VC activity. What to do? Who to tell? Knowing that Coon Skin would most likely be overseeing his favored pastime at that hour, the officer rushed directly over from our

Compound Headquarters. As expected, there was Coon Skin having a grand time.

As soon as he received the details, Colonel Davis ordered, "Let's go, boys." Cards and monies summarily abandoned and left on the table, the Colonel fled to his jeep and upon arrival selected those who would accompany him. Peeling out and leaving the remaining personnel in the dust, he drove full throttle to Can Tho Airfield, jumped into his helicopter, and rescued the team. Yes, this was in the middle of the pitch-black night; and yes, Coon Skin was his official call sign; and yes, he flew his own helicopter. And by the way, he was in the mess hall at 0530 hours, as always, perfectly groomed in starched combat fatigues with a decidedly fastidious comb-over. The very model of a soldier's soldier, he could definitely outrun, outgun, and outperform lesser mortals.

When the date was set for his farewell dinner, Colonel Davis called me over as he passed me on his way to I know not where, and said, "Listen, boy, I don't know what you plan to feed us at this upcoming shindig and I don't really care, but I tell you, boy, I want one of those Alaska things for dessert. You got that, son?"

I made the only conceivable response: "Yes, Sir."

He smiled and said, "Good, that's real good."

It was odd, how being referred to as *boy* and *son* by this man felt like the highest of compliments. Up until that exchange, I had no idea that he had taken any notice of me or had any concept of what I did at Eakin Compound.

While the festivities were under way, I happened to be in earshot when General Desobry, with some evident trepidation, asked one of the other attendees, "Do you have any idea what we're having for dessert?"

He replied, "I believe that Lieutenant Dinan is doing another one of those Baked Alaskas."

The General made no reply but his disgruntlement was clearly palpable. Fortunately, as the General's past reaction was very vivid in my memory, I had decided to create a decidedly different form for presentation than the memorable one I had sculpted on Jay Arnold's behalf. In a way I was surprised that

I was never instructed never to do such a thing again, but I guess the General had more pressing things to think about. At any rate, because of the General's apparent conviction that he was to endure an encore of that event, it was especially rewarding to note the approval and pleasure that erupted when we presented a great, flaming ocean liner inscribed on the port side, "The Coon Skin," and on the starboard, "Bon Voyage." Actually I was taken aback when the assembled made a round of applause and General Desobry gave me a look clearly interpreted as a silent "Thank you."

Do we just imagine that replacements lack the color, the glamour, the dash, the competence, the *je ne sais quoi*, and all those other fine qualities found in their predecessors? Or are we given just reason for making this finding?

During our military participation in the Vietnam conflict, the prescribed tour of duty for all save those officers at the highest command levels was 365 days. Although this is clearly one year, it was never so expressed by those in-country. By the second day in-country, it was usual for one to answer the question, "How much longer do you have to be here?" by responding: "Three hundred and sixty-two days and a wake-up." Turnover of personnel was therefore a very common happening, and in most instances, I had no particular awareness of who came or went, or what their function was. Yet, there naturally were a number of personnel changes of high impact from my vantage point.

FIFTEEN

REPLACEMENTS

COLONEL Davis was ultimately replaced by two colonels. With things heating up, it was determined that the responsibilities of Deputy Commander MACV IV should be split into two jobs: Deputy Commander for Administration and Deputy Commander for Operations. The administrator was Colonel Hill, who was probably the most senior colonel in the army. Colonel Hill was not going to become a general, and he knew it. He arrived quietly, accompanied by no less than eight footlockers filled with booze. He had wonderful dark, bushy eyebrows and an easy manner.

He stopped me one day and asked, "How did you receive your commission?"

I answered, "OCS, Sir."

He said, "I think that's the hardest way; good for you. If you need my help with anything, you'll know where to find me."

The operator was Colonel Fletcher who was probably the most junior colonel in the army. He arrived to great fanfare, insisting that the entire compound be assembled so that we could behold the wonderfulness of his person. He was accompanied by as much military hardware as he could manage to affix upon his over-starched combat fatigues. Included in this display was the Combat Infantryman's Badge; unhappily for him, he had not earned it, and he was soon ordered to refrain from wearing it. All in all, he resembled more a poster than a person, and he was in hot pursuit of what he hoped would be his first of many stars. I will address my encounters with this Colonel in due course.

When Compound Commander, Major Hicks was rotated, he was replaced by Major Magnus, his exact opposite. Where Major Hicks radiated an aura of the perfectly groomed, unflappable, imperturbable, and in-control executive, Major Magnus was at all times nervous, rumpled, and seemingly lost. Many were the times I came across him late in the evening, standing alone at some odd location gnawing on the warts he sported on his right knuckles. He would be sniffing the air, and his hair, even with a military cut, appeared to be disheveled.

He would invariably stop me and ask, "Do you smell that?" Now, as mentioned before, the varied smells prevalent in the Mekong Delta are, for the most part, hideous and to be avoided if possible.

So, the first time I was so queried, I was very much at a loss to understand the actual intent of the question and I answered, "Smell what?"

"Don't you smell it? Don't you smell it? Marijuana!"

I suppose it was there, but I was unable to separate that particular perfume from the abundant competition, so I countered with, "I can't seem to pick it out."

My dealing with the Major was always cordial, although he did seem to suspect that I might be a little crazy. He very likely had good reason for that assessment, and you, the reader, may ultimately concur.

The G4, Commanding Officer of MACV IV Logistics, Lieutenant Colonel Bagley, was replaced by Lieutenant Colonel Bowden. Circumstances preclude my noting anything negative regarding this officer. Suffice it to reflect on the fact that he cut short his tour of duty in the middle of the '68 Tet Offensive by the act of inserting the barrel of his .45 automatic into his mouth and pulling the trigger.

When Master Sergeant Chavez took leave of us, he was replaced by Master Sergeant Erickson. Tall and haggard in appearance, Erickson was an exceptional mess sergeant. He was able to replace Sergeant Chavez so seamlessly that there was no interruption in the flow of operations within the mess halls. When we converted to a Class 1 mess from a COLA mess, he also accomplished this monumental transition without a hitch. He seemed to be on duty 24/7. It didn't matter what the day or the hour; if I showed up, he was there. Until one day he was not there. I was told that he was seeing to some personal business. The next two days followed in kind, and on the fourth day the Sergeant was back on duty. From his general appearance I concluded that his personal business must have been physically punishing indeed. Although he was so much loved and respected by his staff that they all covered for him, I was able to ascertain, in time, that this almost perfect mess sergeant had the misfortune of needing to take care of personal business for three days about every ten to twelve weeks. This personal business attended to by Sergeant Erickson was comprised of consuming so much booze on day one, that days two and three were, of necessity, mandated for purposes of recovery. By the time I figured all this out, without input from him or, for that matter, from my staff, two more occurrences and approximately three times as many months had passed. After giving it a lot of thought and weighing the pluses and minuses, I determined to join in on the cover-up. I knew that I was required to make a report as soon as I became aware of such activities. I also knew what a great job Sergeant Erickson did and that such a report would be his ruin. To do nothing, I needed to maintain a mantle of ignorance, and although some nasty senior NCOs provided me with decidedly pointed hints over time, I simply acted naive—a possible reality to their way of thinking for anyone bearing the rank of a lowly lieutenant. Sergeant Erickson was not a fool, and I did not fool him for one minute. When he knew that I knew, in appreciation of my evident decision to do nothing, he worked even more diligently, although how he managed that, I never knew.

My administrative assistant, Sergeant E-7 Sullivan, was replaced by Sergeant E-7 Butler. Sergeant Sullivan was on the overweight side and approached things with a self-deprecating sense of humor. At the same time, he maintained that aura of professional and life competence that made him something of a father figure among his juniors. In this, I include myself. Sergeant Butler was on the underweight side and approached things with an astonishing earnestness that tolerated very little humor. He had occupied most of his almost thirty years of military service in various administrative capacities. Sergeant Butler said I was the first officer he had ever worked with who actually ran the operation in his charge. Of this he did not approve. He also said that his job as my administrative assistant was the most stressful and difficult job he had ever had.

It was late in the evening when I was alerted by two of my enlisted personnel that something terrible was in process concerning Sergeant Butler. They urged me to go to his quarters quickly. I ran to his quarters as fast as I could and entered without knocking. Sergeant Butler was seated upon his bunk holding his rifle to his head with his thumb in the trigger housing. He looked up at me with tears running down his face and said quietly, "Lieutenant Dinan, I simply cannot take it anymore." Somehow, I made the immediate assessment that a lot of talk would be of little use. So, while he awaited my comments, I approached him slowly as though I was about to say something comforting, and when I was in range I snatched the gun away with all my strength. Sergeant Butler was taken completely by surprise, and this seemingly, and very probably, stupid tactic worked. Disarmed physically and mentally, Butler broke down completely. I held him, and he cried. He held me too, and I cried. I needed someone to hold onto, I was shaking all over and scared half to death.

The following day, Sergeant Butler was taken to a hospital in Saigon for psychiatric evaluation. After three weeks, he was returned to us and finished out his tour without further incident. He was a good man, and he performed admirably.

Every so often when we were alone, he would look at me with the most serious expression and say, "You know I would've done it." I knew.

Getting ahead of myself for a moment, I will mention here that when my year was coming to completion, I made the decision to extend for an additional six months. When I notified my staff of this decision, every one of them did the same. In their own way, they all told me that they had never worked with an officer like me before and that they couldn't miss out on the opportunity to see what would happen next. I'm not certain whether or not this was meant as a compliment, but I took it as such, and I was glad to have all of them with me.

SIXTEEN

MAJOR GENERAL ECKHARDT

THE next major personnel change was to be the replacement of the commanding general. Prior to his farewell, General Desobry took a two week R&R (Rest and Recuperation) holiday. During that limited period, Colonel Fletcher, as the acting OIC, used his temporarily acquired, substantial authority to effect the construction of four private bathrooms attached to the quarters of the commanding general, the two deputy commanders—of which he was one—and those reserved for visiting dignitaries. My intelligence network informed me that, upon his return, the General was disconcerted beyond telling as he considered this augmentation to be a monumental waste of time and money as well as a significant impediment to informal relations between all the members of the General's staff. Evidently it was the relationship or interaction part that most distressed him, but he was leaving, and it seemed that this was something that our Colonel Fletcher needed badly.

My contact with General Desobry was limited to occasions of high-level entertainment, and these included such notables as Senator Edward M. Kennedy, General William C. Westmoreland, Sterling J. Cottrell, who, as DepCORDS for IV Corps, was the senior civilian representative of the U.S. Government in the Delta, and General of the Army Omar N. Bradley.

Sterling was a hoot. Always clad in spotless whites (shoes, knee socks, Bermuda shorts, starched elbow length shirt with epaulettes and IV Corps insignia), he projected the persona of a classic English colonial. His ensemble was completed or augmented by a snappy swagger stick carefully tucked up under his left elbow. He would arrive in his white, chauffeured car, sitting in the rear seat behind drawn shades, and alight only after the driver had opened his door for him. The civilian operatives were supposed to be in the forefront of the war to win the hearts and minds of the non-military Vietnamese population. It was noted that Sterling was probably more of an affront than a forefront toward this lofty goal, but I had no personal interaction with the man. He seemed to me a decent sort who was enjoying one hell of a good time. I did get to deal with the next diplomat who took on the role of the highest ranking U.S. civilian in the Delta, and we will deal with that later on.

Five Star General Omar Bradley arrived at Eakin Compound at midday. As a tall man of obvious senior age wearing civilian clothes, he might have caused little stir as he moved about the compound except for the distinctive five silver stars centered on his baseball cap. The gap between one and five stars is almost inexplicable. This is not simply nobility; this is royalty. As a consequence, royal treatment was provided to the best of our provincial capacity. Lunch was fully prepared and ready to ensure that it could be promptly served, allowing plenty of time for dining and conviviality prior to his scheduled return to Command Headquarters in Can Tho at 1400 hours. A room had been set aside for the General's comfort where he might freshen up in respectful privacy. Availing himself of the lure presented by this comfort zone, the General seated himself in a comfortable chair and promptly went to sleep. As the clock ticked away with no Omar in sight and the quality of luncheon in peril, one senior officer summoned up the courage to ascertain the cause and reported to the assembled, "General Bradley is asleep."

Someone asked, "What should we do?"

I thoughtlessly suggested, "Why don't we wake him up?" After all, my entire luncheon was heading to ruin.

For this indiscretion I received an unpleasant lecture on military courtesy. After all, "You don't simply wake up a Five Star General!"

Time waits for no man, and the ticking clock soon approached and passed 1400 hours. At length, one of the senior officers did simply wake up the Five

Star General. I positioned myself exactly in the line of sight and well within hearing distance to observe what horror might befall this brave soul. Upon being brought to light, the General instantly observed his wristwatch and exclaimed with some alarm, "My God, do you know what time it is? You've let me sleep for over two hours. Don't you know that there is a war going on?"

My radiant smile was clearly observed, if not appreciated, by the assembled. Oh the rigors and rewards of protocol!

Major General George S. Eckhardt was chosen to replace Brigadier General William R. Desobry as the Commanding General of IV Corps. Not wishing to leave any small detail to chance, General Eckhardt scheduled an inspection prior to his actual takeover. Arriving at Eakin Compound from Command Headquarters for lunch, the General's staff and the two Generals assembled in the generals lounge where a special luncheon had been prepared for this important occasion. General Desobry, although junior to General Eckhardt, was the host, and therefore everyone awaited his cue as to how the lunch would proceed. In accordance with his usual practice, General Desobry headed toward the bar in expectation of partaking in his usual pre-luncheon martinis, commenting casually to his superior officer, "Why don't we have a little refreshment?"

The unsettling response from General Eckhardt was "We don't drink midday." Sorrow was evident in all the assembled save Colonel Fletcher who did not drink midday, late day, or any other time of day.

Following the lunch, General Eckhardt made a thorough inspection of the compound, accompanied by the entire General's staff, along with the Sergeant Major of IV Corps, Sergeant Marcille. I was on hand at the completion of this inspection when the entire assemblage was thus, and only thus, instructed, "Your tennis court is not regulation size. When I arrive in two weeks, it will be."

All construction projects in the Delta were put on hold that very afternoon, and all relevant personnel were diverted to Eakin Compound. When General Eckhardt arrived two weeks later accompanied by his tennis gear, his special bed, and several footlockers containing his medical supplies, the tennis court was regulation size.

General Eckhardt was a very senior major general, and although apparently afflicted with numerous physical challenges, he more than compensated for these with his mental abilities. As he led us up to and through the '68 Tet Offensive, he never lost his cool or showed any signs of physical or mental fatigue. Throughout, he faithfully engaged in two sets of tennis daily, not even breaking his usual routine during the height of those hostilities. Always appearing on court resplendent in classic tennis whites with sweat bands on both wrists and about his brow, and armed with a metal-framed tennis racket—at that time, an exotic weapon—the General played a game not of power, but of placement. The prerogative of partnership in this activity was claimed and assumed by none other than our Colonel Fletcher, who isolated the General from social contact with all others save himself to the very best of his ability. The outcome of their matches is unknown to me, but I suspect they may have been compromised by other more pressing goals lodged within our Colonel's breast. Notwithstanding, having watched thousands of hours of professional tennis singles matches, I can report that I have never seen any player who could, with consistency, place the ball with more deadly and devastating accuracy than General Eckhardt.

When our athletic director determined to have a singles tournament, the General, to his credit, threw his lot into the hat. When the lots were pulled, the honor of General Eckhardt's opposition in the first round of the competition fell squarely upon the broad shoulders of an energetic and athletic eighteen-year-old enlisted man with considerable tennis acumen. Upon learning of his privilege and opportunity, this private was moved instantly to a state of blowhard and braggart.

As the club officer, I was possessed not only of the singular opportunity but also the responsibility of presenting myself in the enlisted mens club on a regular basis. As a consequence, I was well positioned to overhear the lengthy dissertation concerning the upcoming humiliation the General was to receive in racket-to-racket combat. The General was to be shattered, smeared, blinded, obliterated, debunked, humiliated, and otherwise humbled in the ensuing engagement. It was a love song in the ears of his listeners. No report was ever given on comments that may have been made by General Eckhardt attendant to the ensuing melee. Me? I had no dog in the fight.

When the great event finally unfolded, the General appeared on court in his usual tennis whites. His enlisted opponent sported attire that may be described as a precursor to the on-court costumes popularized by Andre Agassi. Many were the balls that whizzed by our Commander. His physical challenges precluded chasing down severely placed fast balls. However, every time a ball was placed within the range of his capacity he owned the ball and the court. It was drop balls, lobs, left base line, right base line, up the middle, wherever the opponent wasn't, the ball was. The private persevered, but he could not prevail. Such speed, such dedication, such determination to win must be applauded, but superior skill and strategy won the day. Goliath bested David on this field of battle with a performance demonstrating the result of much-practiced skills as I have yet to see repeated. The David was literally carried from the field of battle having exhausted himself beyond the ability to walk. General Eckhardt, showing more of a glow than sweat, graciously accepted congratulations. In the end he did not win the tournament but he did win our hearts. That was a performance!

SEVENTEEN

COLONEL FLETCHER

A couple of weeks after our Colonel, the new Deputy Commander for Operations at MACV IV Headquarters, had settled in and had been duly relieved of his unearned combat badge, I received a call from his office. I answered this, as every call, in the standard military manner, "Can Tho Mess Association, Lieutenant Dinan speaking, Sir."

"This is Captain Clark. I am Colonel Fletcher's protocol officer. You will hold for Colonel Fletcher."

So, I held and I held, and about two minutes later I was rewarded by this instruction, "Lieutenant Dinan, this is Colonel Fletcher. You will report to my office at exactly 1500 hours."

I responded with the only possible come back, "Yes, Sir."

I arrived at his office complex not later than 1450 hours and immediately identified Captain Clark by dint of the somewhat oversized sign prominently displayed on his desk that advertised Captain Clark, Protocol Officer, MACV IV. As I approached his desk, I was greeted with silence and a very blank stare.

I responded to this warm welcome by giving my name and purpose, "I am Lieutenant Dinan, and I have been instructed to report to Colonel Fletcher at exactly 1500 hours."

The Captain observed me with markedly less interest than that which he showed to his timepiece and, making no comment, left me standing in wonder. He referred to his watch several times, and at what must have been the exactly correct moment, he deigned to address me, "You will go in now," at the same time indicating an open door behind him and to his right.

As I came to the doorway, I could clearly see Colonel Fletcher sitting upright behind his desk, looking expectantly outward, displaying the normal body language of someone expecting a caller.

I walked in and said, "Colonel Fletcher, you wanted to see me." He looked as though he had been struck, as though he could not comprehend the situation at hand. Clearly, I became disoriented at that point.

This feeling did not diminish as he continued to look at me for an interminable amount of time before suddenly inquiring, almost incredulously, "Is that the way you report to a senior officer?"

I said, "You wanted to see me."

He ordered, "You will go back out and report in properly." Retreating to the far side of his doorway, I caused three distinct knocks on his open door. Following a distinctly rude lapse of time I received my summons, "You will come in."

In I came and I snapped to attention exactly on the square in front of his desk. While rendering my most military salute, I proclaimed, "Lieutenant Dinan reporting as ordered, Sir."

In the finest tradition of ball-breaking, I was left standing thus for a full one-minute count before my salute was casually returned and the instruction given, "You will sit." This was accompanied by a distinct direction toward the chair on his right. I took that instruction immediately, but I have pondered the consequence should I have opted for the chair on his left.

Following this inauspicious beginning, the Colonel made several observations and comments that seemed to be of little, if any, import. I was sitting up very straight and looking at him most directly, displaying my utmost attention. This was not to be enough.

Colonel Fletcher interrupted his monologue to instruct me: "Lieutenant Dinan, when I talk, you will take notes." I had a pen, but lacking a proper pad, I utilized the back of my chit book (you remember those little books we purchased for the purpose of paying for drinks) to record his important remarks. This offense did not escape the Colonel's keen observation and he, with some joy, inquired, "What is it that you are writing on?"

Said I, "A chit book."

He responded with apprehension, "A chit book?"

I confessed, "Yes, Sir. A chit book."

At this point I became the beneficiary of this instruction: "Lieutenant Dinan, whenever you report to a senior officer you will have a proper pad and writing implement on your person; is that understood?"

"Yes, Sir." What else could I say?

When the interview concluded with my dismissal, I went away wondering what the point of this meeting might have been. The only thing that came to mind was that I was being alerted to the proposition that the Colonel was a person who ought not to be trifled with. Since trifling with any colonel was not on my agenda, the whole thing made little sense to me at that time.

Over the next several weeks, I was the proud recipient of numerous calls from the Deputy Commander's office. The routine was steadfast. I would answer my telephonic summons, "Can Tho Mess Association, Lieutenant Dinan speaking, Sir," and the unannounced but recognizable voice of Captain Clark would instruct me, "You will hold for Colonel Fletcher." My holding was significant; the ensuing exchanges were not. I was just beginning to seriously wonder why in the world I was receiving these senseless interruptions when the absolutely unexpected occurred. I answered my telephone and was shocked to be directly confronted with a voice announcing, "This is Colonel Fletcher."

Summoning all my military training, I was able to exact the cogent response, "Yes, Sir."

"Lieutenant Dinan, I need a coffee pot for my office."

"No problem," I responded. "I'll send someone over to the PX and I'll have the purchase charged to your mess account. Do you wish this to be delivered to your office or to your quarters, Sir?"

"Lieutenant Dinan, this is not for me; this is for my office."

"Yes, Sir?"

"Doesn't the mess association have a coffee pot that could be put to use in my office?"

"No, Sir. We have only fifty-gallon coffee urns and no extras that I know of."

"Lieutenant Dinan, this is Colonel Fletcher, and I believe that there must be a way you can accommodate the needs of my office. Do you understand that?"

About this time my Bronx Irish heritage began to inflict itself on my thought processes, and I thought to reply, with all due military courtesy, *Screw you, you cheap son of a bitch, you've been intimidating the wrong boy.*

I answered, "Golly, Colonel Fletcher, you know that I want very much to accommodate the needs of your office, but how can I justify using the funds provided by our enlisted personnel to provide conveniences for our senior officers?"

"Lieutenant Dinan, this is Colonel Fletcher," as though I had forgotten the identity of my caller, "and this is not for me; this is for my office."

A salient point I suppose, but I had a difficult time reconciling the difference. Besides, my Irish was up, and I'd go to hell before I'd lay me down. So I stalled and said, "Gee, there must be some way that the mess association can be of assistance, but I am stymied as to what that way might be."

With a distinct change of pitch that registered as a triumph in my tender ears, I was treated to the response, "Of course, Lieutenant Dinan, there is always a way, and it is up to you to discover and to implement that action."

Now I was burning. If smoke could have come out of my ears, I would have set off an alarm. Although shaking, I contained myself as a thought of particular appeal entered into my thinking, "Colonel Fletcher, I think I may just have an idea."

"Yes, Lieutenant Dinan, yes." Oh, the victory in his voice, you could embrace it. "What is it that you have in mind?"

"Well, Sir, you know that we have monthly board meetings of the Can Tho Mess Association. I propose that at the next regular meeting I will recommend that the mess association purchase a coffee pot for your office and just as soon as the General approves the minutes of that meeting, a coffee pot will be purchased and delivered to your office."

The response was delivered in a voice so cold as to be numbing, "Lieutenant Dinan, this is Colonel Fletcher; can you help me or not?"

As I was too hot to be cooled off, I sweetly responded, "Colonel Fletcher, this is Lieutenant Dinan, and I have given you my best offer."

How is it that you can hear the difference between a handset being placed or slammed onto the holding unit? This was a slam, big time, and the Colonel seemed to lose interest in me. No calls, no intimidation, no nothing. Captain Clark purchased a coffee pot for the office out of his own pocket. Remember, it was for the office and not for Colonel Fletcher.

Following that frightful exchange, Colonel Fletcher and I had no contact until one sunny Sunday morning around 1100 hours. I had just left the mess hall and was walking diagonally across the parking lot when I noted a bit of a commotion outside of the snack bar. In front of the doorway were two tall Americans clad in tennis whites, holding rackets and balls. The one maintaining a calm demeanor, I immediately recognized as General Eckhardt. The one that was screaming, all red in the face with bulging veins, was our Colonel Fletcher.

Evidently, following tennis, the General decided to make a tour of the compound and at that time he was seeking entry into the snack bar. Inside was a fifteen- or sixteen-year-old Vietnamese lad who was known as Boy-san. His purpose for being there was to clean the place prior to the opening at 1200 hours. He had strict instructions never to let anyone inside other than those he recognized to be duly authorized. These two in tennis whites he did not recognize.

I could clearly hear the Colonel who was screaming at the top of his voice, "You will open this door immediately, *immediately*! Do you understand me?"

I slowed my pace in order to prolong the enjoyment, and I could distinctly hear Boy-san politely explaining, "*Thieu Uy* Dinan say no."

"Dinan," bellowed the Colonel, "Dinan! Goddamnit, you will open this door now. I'm Colonel Fletcher, and you will obey me, do you hear?"

Boy-san held his ground and repeated the mantra, "*Thieu Uy* Dinan say no." I was proud of that young man.

When I got within a few feet of this marvelous exhibition, I came to attention, saluted, and gave a verbal greeting to both officers, "Good morning, Sir. Good morning, Sir." After the General returned my salute, I greeted the besieged, "*Mon joi*, Boy-san."

At this juncture, the Colonel turned on me, red in the face with bulging eyes and veins ready to pop. The poor chap could hardly speak. "You, you, you will, you will instruct this boy to open this goddamned door."

I responded while pretending not to notice his fury, "Of course, Sir, my pleasure, Sir. Boy-san, please be good enough to open the door."

I could see that the lad was half-scared to death, but he smiled up at me, said "*Mon joi, Thieu Uy* Dinan," and promptly opened the door.

As we stepped inside, Colonel Fletcher let me know in no uncertain terms that "This boy should know who I am, and when I tell him to do something, he had better do it." I said that I was sorry for the confusion but that considering how they were dressed, Boy-san had no way of knowing their status.

To this, the Colonel responded that the way he was clothed should be of no consequence and that under any and all circumstances, "This boy should know me and know me well; you will see to it."

So I told Boy-san, "This is Colonel Fletcher. Take a good look at him and remember who he is." The lad looked at him, evidencing little comprehension, but he was shaking his head up and down using the universal body language for *yes*. So I continued, "Whenever Colonel Fletcher asks you to open a door for him, please open it."

More head shaking accompanying a vacant look, but the Colonel seemed to be pleased, and as he looked about he declared, "This place needs to be cleaned."

I said, "Except for the interruption, that is exactly what this young man would be doing."

General Eckhardt was obviously enjoying this entire episode, and he barely suppressed a serious guffaw that did not go unnoticed by Colonel

Fletcher, who tried to recover the situation by ordering: "See to it that it is cleaned thoroughly. I will be back to make an inspection."

I provided him with the requisite "Yes, Sir" and out the door he went. The General paused long enough to give me a warm smile and, saying nothing, left also.

EIGHTEEN

MORE ON COLONEL FLETCHER

COLONEL Fletcher did not complete his tour of duty as Deputy Commander; instead, he was given command of a battalion somewhere, presumably for his career benefit. His replacement was in the person of Colonel Hazen, generally referred to as "Shazam." I'll give him proper introduction later on. I don't know what became of Colonel Fletcher, but I do know that when helicopters became in short supply, he was the first to have his reassigned by his former staff at Command Headquarters. Colonel Fletcher was a perfectionist. This is in many ways a strength, but when taken too far, it can be a weakness. Colonel Fletcher took it too far, and in doing so exhausted those who reported directly to him.

Just a couple of examples will serve to demonstrate this unfortunate result. The five- and six-man teams out in the field were not issued typewriters because typewriters were not in their TOE (Table of Organization and Equipment). Notwithstanding this fact, Colonel Fletcher issued orders requiring all their reports to be submitted typewritten. But not only this. He further insisted that they be letter-perfect. Somehow the teams were able to find typewriters, but alas, they were unable to find clerk typists. Typing attempts by non-typists working out in the field under inconvenient circumstances tended to fall short of perfection. Upon receipt of a less than perfectly executed report, the Colonel would refuse to process it. Instead, he would send it back to the originator with redo instructions as many times as necessary until he was in possession of a perfect copy. This had the teams screaming and howling "Unfair!" but they were led by either a major or captain, and the boss was a colonel.

At one point the Colonel had a major report consisting of no less than forty-seven–single-spaced pages, which was destined for Headquarters Saigon. Clearly it was required that this report be letter-perfect. In the days before word processors, 100 percent letter-perfect, meaning no corrections of any sort, on forty-seven–single-spaced pages was a monumental undertaking of Olympian proportions. As it happened, a senior lady operative in Sterling J. Cottrell's civilian corps was known to possess this unusual ability. Under great pressure, she submitted to the request and undertook this monstrous task, which had to be accomplished in very short order. Time was of the essence, and she dutifully stayed up typing all around the clock for two days. In an exhausted state, the lady triumphantly presented the letter-perfect, forty-seven-page report directly into the eagerly waiting hands of Colonel Fletcher. Neither rising from his chair, nor making any greeting whatsoever, the Colonel placed the volume upon his desk. He then reached into one of his desk drawers and produced a ruler. It must be noted at this point that the military establishment had just recently made a change to their formatting rules. To wit, what was yesterday's requirement, namely, a three-quarter inch margin, was today a full one inch margin. This important intelligence had not yet been made known to Cottrell and his cohorts, and as a result, when the ruler was laid to paper, a deficit of one-quarter of an inch in the margin was clearly measured. Handing the report back to the lady, the Colonel addressed her for the first time with this comment, "The margins are incorrect. This must be retyped." The lady replied with some very unladylike remarks and, additionally, some very unladylike suggestions as to certain things she felt the Colonel should accomplish personally utilizing the report in question. Then she not so gently slammed the report to rest on his immaculate desk, walked out, and went home to bed. I don't know what subsequently transpired concerning this report, but I am certain that when it was ultimately sent to Saigon it was perfect in every military way.

Colonel Fletcher's operating style did impact me directly as a result of two directives. One directive created the position of "saluting officer." In the pursuit of a more refined level of military courtesy, it was ordered that on a twice-daily basis, a young officer, i.e., a lieutenant, would place himself in front of headquarters in a position convenient to engaging any and all as they entered or left Eakin Compound and, from this vantage point, give and take salutes. The times were set at 1200–1300 hours and 1700–1800 hours. Since there were but three lieutenants to be found, we three defended the art of the U.S. Army salute many hours per week, basking in the South Vietnam sunshine or swimming in the rain, all the while being blessed by the dust or mud kicked up by the jeeps as they rounded the corner into the parking area. Most people looked upon this activity as a joke, and I suppose that, in its own way, it was, except the joke was on us.

The other directive that was to exact a toll upon my tranquility was that the monthly minutes and financial statements of the Can Tho Mess Association were to be fully staffed prior to submission to the Commanding General (CG) for his signature. What this meant was that each head of operations, the G1, G2, G3, G4, and the IG, would be required to review and initial these monthly reports personally before they were delivered up to the CG. As you may guess, most of the time these gentlemen had little knowledge or understanding of what was put before them, so it became my duty to answer each and everyone's particular questions. This situation led to a certain amount of friction that in due course led to confrontation. I will elaborate on this in the next chapter, but first I must conclude my narrative on Colonel Fletcher with one of my fondest memories.

The Colonel was still ensconced at Command Headquarters when we began a major upgrading of the officers club facilities. The furniture and fixtures that we determined we wanted were not available in the Vietnam marketplace. Consequently, I was called upon to engage in a major shopping adventure in Hong Kong. Actually, I had to make the trip twice. On the first trip, I utilized my personal R&R allowance, and on the follow-up trip I was sent, under orders, for business purposes. The key thing about shopping in

Hong Kong for government-related purposes at that time was that one had to ensure that the monies spent would go into the hands of approved merchants only, lest our U.S. money should wind up in the hands of the Communists. This could be tricky, and many American heads rolled for the folly of being fooled by seemingly benign and benevolent Chinese merchants who asserted their approved status, bolstering their claims with bogus documentation.

Having unloaded about $25,000 in actual U.S. dollars in that fabulous city, I returned triumphant to Vietnam. I was expecting a hero's welcome when I reported to the Can Tho Mess Association Board of Directors and regaled them with the details of the glorious bounty that should soon be ours. It was a severe disappointment to learn upon my return to Eakin Compound that my first order of business would not be to report to the board, but rather to report to Colonel Fletcher. The Colonel, coincident with my departure, had strict orders delivered to Sergeant Butler to the effect that my very first act upon my return would be to call his protocol officer, Captain Clark, and schedule a meeting with the Colonel himself. This, of course, I did. A meeting was duly scheduled, and I reported as ordered.

Arriving at the door of the Colonel's office, I once again caused three distinct knocks and entered when given instruction to do so.

Coming to attention in front of his desk and raising a very smart salute, I announced, "Lieutenant Dinan reporting as ordered, Sir." To my utter surprise, Colonel Fletcher seemed to be actually happy to see me.

He returned my salute smartly, stood in welcome, and very cordially offered, "Lieutenant Dinan, please be seated and make yourself comfortable."

I did so, and as my superior continued to exude copious expressions of joy, warmth, and happiness, I was gripped by the sense that something was wrong with this picture. The joy, the warmth, the happiness—they all proved to be genuine emotions, but these fine feelings were, although outwardly apparent, internally directed. The Colonel was happy for himself, thinking that he had me by the balls!

"We have a serious situation here, Lieutenant Dinan." He looked upon me with a warm smile awaiting my response.

I said, "Yes, Sir. What may that be, Sir?"

Taking a deep breath and sporting a look of great concern, Colonel Fletcher took a few moments and then proceeded, "I believe that you have been on a buying trip to Hong Kong to make extensive purchases on behalf of the mess association. Am I correct, Lieutenant Dinan?"

"Yes, Sir. You are."

Giving a nod that communicated *of course I am,* he continued, "My problem with this is that buying in Hong Kong is a very serious undertaking. You, young and inexperienced, may very well have engaged in acts considered treasonous." Although he paused here, allowing time for me to vigorously defend myself against this absurdity, I had not been asked a question, and I opted for silence.

Once he became convinced that I was not going to offer a response, he proceeded: "You may not be aware of it, but if any of your purchases were made from an unauthorized seller, it will be my duty to bring you up on charges and have you court-martialed. Do I have your undivided attention now, Lieutenant Dinan?"

As the light was beginning to shine on the cause of his joyful self-satisfaction, I replied, "You always have my undivided attention, Sir."

He got the point, and giving me a rather cold look, he continued, "You will provide my office with a detailed report specifying each purchase and the company from which the purchase was made; is that clear?"

"Yes, Sir."

"Good, Lieutenant Dinan. I will have each one investigated, and if just one is not specifically noted on the approved list, I will bring you up on charges. I don't think you realize how serious this matter is." Once again, he paused for a response. Once again, I failed to respond. And once again, accepting that I would not make a response to a statement, he forced my position by inquiring, "Have you anything to say, Lieutenant Dinan?"

Why does he think it is necessary to keep reminding me of my name and rank?

I said, "Yes, Sir. I think I can save you a lot of time and trouble."

With this his eyebrows jolted up so far and fast that I was truly astonished at the spectacle. "Oh, you think so, do you? And just exactly how do you propose to save me a lot of time and trouble, Lieutenant Dinan?"

The God of good fortune smiles upon each of us from time to time, and I remain especially grateful for this particular blessing. Shortly prior to my sojourn, Colonel Fletcher himself had made the same pilgrimage. He had gone in quest of air-conditioners for use in the quarters of such senior officers as himself. On this self-serving mission, he had directed all of his purchases through the offices of one Mr. Lee Sang Wong. On my mission for the mess association, I had done likewise.

I, affecting that country-boy demeanor that everyone hates, said, "Golly, Colonel Fletcher, I was sure that I'd done right when I did my buying through the very same agent as you."

With great surprise he inquired, "Lieutenant Dinan, what are you talking about? Explain yourself."

I could see the veins beginning to come up, but I was on a roll, so I continued theatrically, "Why, I'm talking about Mr. Lee Sang Wong; what a nice gentleman, and he says that I should tell you what a great pleasure he had in dealing with you."

The eyebrows that had so vividly gone up now came crashing down, and I was assaulted by two very cold eyes violently engaged upon my person. He spoke, "Where did you learn of this? Who told you? I demand to know."

Keeping in character, while pretending to be unaware of this cataclysmic mood swing, I replied, "Golly, I just don't know. You know that when I considered that I would be going all the way to Hong Kong and that I didn't know anyone there or anything, I kind of figured that I should get me some real good advice. So I asked around and around, and everyone was real helpful like, and one feller mentioned, golly, I wish I could remember who he was, that you had done all your dealings with a Mr. Lee Sang Wong. Hearing this, I figured that with me being so young and inexperienced and all, you know, that the smartest thing I could do would be to do the same like an experienced officer like yourself did. So I found this feller Wong, and he became my main man."

Colonel Fletcher was now looking at me with an expression of incredulity. The gas had escaped from his balloon, the joy had fled, but he recovered sufficiently to inquire, "Can you prove that all your purchases were made under the direction of Mr. Wong?"

I answered, "My report with purchase receipts and other documentation will verify my assertion, Sir."

At this point Colonel Fletcher left me hanging while he made a meticulous inventory of his ceiling, and after being brought back to the reality of what was happening on the ground by this technique, he in a rather deflated manner addressed me, "Lieutenant Dinan, this interview is over, and you are dismissed." That's military speak for "Leave my presence now, immediately, without comment."

I responded in the usual military manner by standing to attention, bringing up a salute, and declaring, "Yes, Sir. Thank you, Sir."

As I approached his doorway, I was stopped in my tracks by the simple words "Lieutenant Dinan."

I turned, "Yes, Sir?"

"You will forget about this meeting, and your report will be unnecessary. Is that clear?"

"Yes, Sir."

Upon returning to Eakin Compound, my first impulse was to go into the PX, buy a coffee pot, and have it gloriously gift-wrapped and delivered to Colonel Fletcher. My second thought was to simply be happy for my good fortune and do nothing. Nothing is what I did, and in so doing, I chose the best option. It is compelling to consider what would have transpired if I had provided the coffee pot in the first place.

NINETEEN

THE ELABORATION

IN a sense, the Can Tho Mess Association was a business, and I was the chief operating officer. As with most businesses, there was a board of directors. My board, or the board, consisted of six U.S. Army personnel: three commissioned officers and three non-commissioned officers. The commissioned officers included a lieutenant colonel, who acted as the chairman of the board, a major, and a captain. The enlisted members included a master sergeant E-8, a sergeant first class E-7 or a staff sergeant E-6, and a sergeant E-5. Regular board meeting were convened monthly, and I would read the minutes of the previous month's meeting and present the current month's financial statements. As with all ongoing business ventures, we were required to operate in the black; however, not so far into the black as to give rise to the idea that we were more successful at making money than in providing service. Once the financial aspects had been fully discussed and agreed upon for submission, the board members would offer up their ideas as to what should be done to better serve our constituents. As this was not truly a professional board of directors, it is hardly surprising that any number of the resulting brainstorms, if enacted, would serve only to muddy the waters. As COO, it was my duty to explain to the various members just exactly why their ideas were unsound from an operational standpoint. Sometimes I prevailed, and sometimes I did not. One instance where I failed to prevail will shed light on my plight.

With the approval of the board, I had secured a soft ice-cream machine and installed it in the snack bar. The product was very much like what you would find in a Carvel's or Dairy Queen, and we were able offer for sale both chocolate and vanilla ice-cream cones to all comers. This proved to be a very popular and sought-after item of purchase. The cost per serving was computed to be between eleven and twelve cents and the price per purchase was set at fifteen cents. So far, so good. However, it was to be announced at one of our regular meetings by our most junior enlisted board member that he recalled hearing someplace or somewhere that no less a personage than President Kennedy himself had stated that an ice-cream cone should never cost the purchaser more than five cents. For unfathomable reasons, this engendered the approbation of the assembled board members. I was very much against this, since losing seven cents on every sale of a popular item would quickly turn our slim margins from black to red. However, no argument I advanced could dissuade the assembled, and with a unanimous vote cast in favor of the five-cent cone, I was ordered to adjust my price list accordingly. Would you believe that the machine almost immediately became inoperative, and although we were unable to offer a five-cent cone, we did remain in the black? Happily, after an appropriate breakdown time lapse, we were able to reintroduce the cones with no fanfare, no objections, and no price reduction.

Along with serving as the whipping boy at these meetings, I also served as secretary, and in this capacity, I was responsible for writing up the minutes that would be attached to the financial statement and submitted to the very highest level for signature, prior to being archived forever. When I say the highest level, I am referring to the Commanding General MACV IV. Although it may seem that this would represent too trivial an item to engage the concern of someone with so very many pressing responsibilities as the leader of all U.S. military personnel in the Mekong Delta with a war in progress, it was nonetheless the way the system worked. It is part of the strength of our military that the buck does indeed stop only at the top.

As noted earlier, these sacred papers were fully staffed before alighting upon the CG's desk. When any of the several senior responsible officers would come across an item in the minutes that revealed such foolishness as a board decision to sell ice cream at a loss because of a Kennedy recollection on the part of one of our enlisted men, they would find that in good conscience or at least in good military practice, it would be imprudent to add their signature of approval in the vetting process. I would get the call, "Lieutenant Dinan, we cannot send this to the General." Thus would begin the process of reconstructing the story line to reflect a decision-making process that would be palatable to the CG and to whoever else might review these documents in the future. Few indeed were the minutes I submitted as a recording of the actual happenings that did not require substantial convoluting to achieve an acceptable narrative. As a consequence, I spent many hours each month reporting to the many G's at MACV IV Headquarters. As a consequence of this latter duty, I made a tactical error that was to cause me much tribulation.

I answered my phone, "Can Tho Mess Association, Lieutenant Dinan speaking, Sir." I was not comforted by what I heard:

"Lieutenant Dinan, this is the Inspector General, Major Nordquest. I want you to come to my office immediately to discuss your financial statement."

I replied, "Yes, Sir. I'll be there as quickly as possible."

It was just after lunch, and it was one of those glorious days of scorching sunshine and breathless air on which I had done duty as saluting officer. I was out of comfort and in foul humor but, as ordered, I got myself to Command Headquarters as quickly as I could and sought out the sign advertising MACV IV IG. This office was not air-conditioned, and in an attempt to prod a little fresh air into the place, the door stood open. Immediately to the right was the Major's desk, behind which sat Major Nordquest and upon which my financial statement was prominently displayed. Along the opposite wall were a number of chairs, one occupied by Lieutenant Colonel Short, who I knew to be the MACV IV G2 (Intelligence). I did not recognize the Major, so I figured that he had to be newly installed, but the nameplate on the desk clearly identified the occupant.

I came to attention in front of the desk, saluted, and announced,

"Lieutenant Dinan reporting as ordered, Sir." After he returned my salute, I acknowledged his guest, "Good afternoon, Colonel Short."

The Colonel was rather easy-going and he welcomed me with, "Hi, how're you doing? You made good time getting here. Now see if you can answer Major Nordquest's questions."

I simply replied, "Yes, Sir," and gave my attention to the IG.

He looked at me rather blankly and began, "Lieutenant Dinan, I have before me your financial statement. I know about these things, and I can see that we have a problem here."

To this intriguing opening I responded, "What may that be, Sir?"

After giving me a sharper look he said, "What the problem *is,* not *may be,* is right here in the first column. This column is designated as Warehouse and in the final entry you indicate a value increase. The only way a warehousing facility can show a value increase is if you reprice your inventory at current prices that are higher than the price you paid upon purchase. This accounting method is not allowable in the military, and you will have to reorder your figures."

I figured that this guy had a clue, and setting the record straight would be no problem. I said, "I am aware of that, and my entire inventory is carried at purchase price. If you look down the column just below Issues, which is a minus figure, you'll see Sales, which is a plus figure. This represents the sales we make to the various teams in the field that we support. According to regulations we are entitled to charge 5 percent above our cost when making these sales, and we do so. This is why this activity shows a profit."

Going back to the blank look, Major Nordquest replied, "Lieutenant Dinan, I don't know what you are talking about, and I don't think that you do either. Are you telling me that you are charging a mark-up on groceries sold to your fellow American soldiers? Because if you are, I think that you're in way over your head, and what you've been doing is not only immoral but illegal, and someone is going to have to pay the price, and that someone is going to be you."

Certainly, I was taken aback; I had to make a readjustment in my thinking.

Now I figured that this guy did *not* have a clue, and that I would have to walk him through this slowly. I could see out of the corner of my eye that Lieutenant Colonel Short was paying very close attention, so as I addressed the Major, I watched the Colonel also to see his reaction.

"I know that this may be a little irregular, Sir, but the regulations clearly state that one mess association may sell to another mess association in an emergency, and when they do so they are entitled to charge 5 percent over their purchase price to cover the costs associated with buying the goods in the first place. We currently sell to about eighty of the five- and six-man teams stationed out in the field (ultimately we had a client roster of one hundred and eight teams), and although they are not officially mess associations, they are de facto mess associations in that they are all on COLA, they pool their monies, and associate for the purpose of getting fed. What we do is send people to the PX in Saigon to purchase substantial quantities of produce in bulk, then resell the items in small quantities suitable for small teams without refrigeration or proper storage facilities.

"By providing a grocery here at Eakin Compound, we save these teams many man-days per month that they would have to expend on buying trips to Saigon. Further, we save them a good deal of money outright because the costs of transportation, lodging, meals, and other expenses incurred in these Saigon trips can be very substantial. Although it does cost us a great deal of time, effort, and money to provide this support, the teams really appreciate it, and we do manage to stay in the black in spite of our basic costs and the spoilage that regularly occurs."

I could see that Lieutenant Colonel Short had taken this all in and he was nodding in approval. Unfortunately, the Major was experiencing some difficulty in grasping the situation.

He said, "I don't know anything about these regulations you are referring to, and I think you are feeding me a lot of bunk."

The Colonel interjected, "Major Nordquest, I am familiar with the regulation that the Lieutenant is talking about."

The Major said, "Thank you, I'll have to look into that, but in the meanwhile we still have this problem with the warehouse showing an increase in value." At this point I was somewhat astounded. I thought that I had just finished explaining this. The Major looked to me and said, "Well, Lieutenant Dinan, what is your explanation?"

So I explained again, and again, and yet again. During this entire time, I had not been offered a seat, so I remained standing, and it simply did not matter what I said or how many ways I explained the financial statement; the reply was consistent: "I still don't understand how a warehouse can show an increase in value."

This went on for a good two hours and I was at my wits' end. I noted that the Colonel was shaking his head in disbelief. I knew that he had understood immediately, and I could see that the Major was stuck in such a way that he was never going to understand. In a state of total frustration I blurted out, "I think that we are wasting our time here."

"What is it that you are saying, Lieutenant Dinan, or should I say, what is it that you are trying to say?"

I lost it and replied, "What I am trying to say, Sir, is that you will never understand what you are being told."

"And, why is that Lieutenant Dinan?"

And here is where I completely lost my military self-control: "Because you are stupid, Sir."

Upon hearing this, Lieutenant Colonel Short completely lost his military composure and with uncontrollable laughter actually fell out of his chair and onto the floor, clutching his sides for all he was worth. This scene visibly troubled the IG, and he was somewhere between angered and bewildered.

He looked up at me in wonder and in the best military tradition took control of the situation by announcing, "That will be all, Lieutenant Dinan. You are dismissed."

On cue I came to attention, offered a salute and said, "Thank you, Sir." Immediately I knew that I had committed a grave error but since it was too late and getting later, I did the only sensible thing: I returned to Eakin

Compound and repaired immediately to the officers club for a couple of soothing scotches.

This had not been an especially good day. I knew very well that I had deported myself foolishly, falling far short of demonstrating ideal military decorum. I figured that although I had been responsible for humiliating Major Nordquest in front of a higher ranking officer, the fact that I had responded to a direct question from a senior officer, and included the respectful "Sir" in doing so, precluded the possibility of insubordination charges. At the same time, I also understood that some retribution was inevitable. My thinking was correct, yet over the ensuing fourteen days I was lulled into a complacent attitude regarding the entire situation. In fact, my financial statement, along with the attached minutes of the board meeting, had been signed and approved by the General without any further deliberations that required my presence. The case was closed, over, and done with. Well, not exactly. My breach of military decorum could not go unanswered—and the answer was to come in the form of several applications of the numerous investigative tools placed in the hands of the Inspector General.

TWENTY

THE HUMILIATION

WITH lunchtime quickly approaching, I was diligently working at my desk. Into my office walked two strikingly tall U.S. Army captains, under arms, and without hesitation, they approached my desk. As I looked up, one of these warriors looked down and demanded, "Are you Lieutenant Dinan?"

When I acknowledged the truth of that circumstance, I was instructed, "You will stand up and remove yourself from your desk. You will not touch anything; you will not speak to anyone; you will not return to this office until you are ordered to do so." Naturally, I was intimidated and bewildered, and this was not lost on these combatants. At this juncture, I was presented with some papers and told, "These are our orders. You will read them after leaving this office, and we will call upon you when we deem it appropriate." These fellows knew how to be truly intimidating; as one of them removed his weapon from its holster, I quickly determined that exiting was my best option.

Upon returning to my thankfully air-conditioned quarters, I instinctively constructed a noble noonday martini in order to reposition my tranquility, and after engaging some opera highlights on our sound system, I seated myself upon our comfortable sofa and began reading the orders so ingloriously presented to me. Retribution was showing its ugly face. The essential gist of the documents I was presented with told the story that the MACV

IV IG had concerns with respect to the financial integrity of the Can Tho Mess Association, which was operating under the currently assigned officer, Lieutenant Dinan, and a full investigative accounting audit was recommended.

Opportunities for fraud and theft for anyone in my capacity were multiple and compelling. Happily, I had made the decision at the very beginning of my tour of duty to maintain a "Caesar's wife" position of being beyond reproach in all my dealings. There was no instance wherein anyone, in any way, could demonstrate that I had abused the operation under my charge for personal gain. Armed with this comfortable knowledge, I used this time of exile not for the purpose of worry but for the far more glorious and noble purpose of enjoying the very greatest of martinis and music. In a way, although initiated under the most ignoble of circumstances, this was an ideal on-post vacation.

On the third morning following my humiliation, I was in bed reading when I heard a crisp knocking at my door. I answered rather loudly to overcome the sound of the air-conditioner: "Come in."

In came one of the combatants, devoid of weapon and notably unsettled by the unexpected opulence of these quarters as well as my unanticipated relaxed attitude.

"Lieutenant Dinan, you don't seem to be very concerned about our investigation."

Music was playing softly, flowing through our Pioneer speakers from a tape rolling on our TEAC A-4010S tape deck supported by a Sansui amplifier, and the air-conditioner was humming merrily, defying the outer temperature, which was approaching 100°F with humidity measuring well into the high 80s. I, ensconced in my bucolic bed, cigarette in hand, and sipping from a cup of hot coffee, invited this bewildered official to help himself to a cup from my thermal pot that was always freshly filled and found on a tray with cups and saucers directly adjacent to my bed. At the same time, I allowed that he might refill my cup also. Not knowing what else to do, he did both.

Finally, seated somewhat comfortably on the bunk across from me, he began: "You know—your books are unbelievable. We have never seen anything like them during our military tour."

It was something between a statement and a question, and as it seemed to be on the complimentary side, I responded, "When I took over, the financials were of a sort that made no sense to me, so I simply changed them into the standard double entry accounting system used in the hospitality industry."

He replied, "Well, we are impressed. Get yourself dressed and come to your office, and we'll go over some questions we have concerning your books."

I asked, "Is there anything I should be concerned about?"

"If there were, Lieutenant, I wouldn't be sitting here having a cup of coffee with you. I would have come with the military police and placed you under arrest." An *arresting* thought. I thanked him, offered more coffee, which he declined, and it was agreed that I would report to my office as soon as possible.

He left. Vaulting out of bed with a confident sense of well-being, I proceeded to engage in at least two of the three required day-beginning rituals. I showered and shaved. Donning my smartly starched and pressed jungle attire, I reported directly to my office. Upon arrival, I noted that the two accounting captains had completely overtaken my desk, now laden with reams of paperwork. Although it had been established that I had not committed any outlawed activity, I was nonetheless apprehensive regarding "some questions we have concerning your books." So, as I approached this team with some trepidation, I was particularly encouraged by their exuberant and warm welcome: "Yo, Lieutenant Dinan, come over here. We need clarification on a few specific entries." In one way I was on guard because I realized that these folks were sent out after my flesh rather than my friendship, but on the other hand I had just been told that I had been found to be fraud innocent. Thus encouraged, but still apprehensive, I sought to first have my position as an innocent reaffirmed.

"Listen," I said, "if there is anything questionable that you've come across, can we get that resolved before we delve further into specifics?" I was happily surprised by the response.

"Listen, Dinan." I was already disarmed by the omission of the lieutenant reference. "We know that you are on edge and on the defensive and that we probably terrified you in the manner of our initial confrontation. The fact is, when we are called in, someone always goes to jail." What a *comforting* thought. "But in this case, we are trying to figure out why we were called in, in the first place. Any accountant could easily determine that there are no inconsistencies in your books and that no theft is transpiring. So we have to ask, 'Why are we here?' There must be something for us to find."

I said, "Well, I did tell the IG that he was stupid, but beyond that I don't know of any other wrong that I have committed."

This confession gave rise to a response of convivial hilarity buttressed by the declamation, "Now we know why we're here. For someone who can keep such smart books that was a particularly un-smart observation to verbalize."

Acquiescing to the self-evident truth of this observation I proceeded to inquire, "When you say 'any accountant' I figure that you must be some special type of accountant, but I don't understand what that could be." Soon I was to receive a lesson I will never forget.

Scanning the office, the lead accountant captain noted an unwelcome, if not unusual, attention to our dialogue. Looking at me with a discernibly happy smile plastered on his face, he recommended, or rather instructed, "I think we should continue this conversation in a more confidential arena."

I suggested my quarters as a suitable venue, and my recent visitor vigorously concurred. I suspect that the air-conditioning had much to do with his enthusiasm for this universally agreed upon proposal.

Upon entering, the joy enveloping my guests was not only evident, but well expressed by the simple comment, "This is unbelievable." While engaged in the frivolous but compelling activity of activating our sound system, I, assuming the role of quintessential host, made inquiry as to my guests' libation preferences. We settled upon manly drinks; I'm talking about seriously refreshing adult beverages. Thus luxuriating in air-conditioned splendor with soft music, soft seating, and hard drinks, I was educated about an investigative occupation of which, theretofore, I hadn't even the slightest knowledge.

My seemingly senior, in-charge-of-the-investigation guest, now nestled comfortably and coolly within my domicile, proceeded to educate me:

"We are forensic accountants. We worked together in civilian life doing mostly corporate work. A company would suspect or actually know that someone within their organization was stealing from them, but they could not figure out who or how. That's where we came in. By tracking their records of debits and credits using numerous formulas we had developed, we were able to pinpoint exactly where the theft was taking place. Believe me when I tell you there is no place to run and no place to hide. No matter how convoluted or clever the construction of the cover-up, the numbers never lie. If you know how to look, and we do, the facts will always be found.

"We had a really good business going, and when I got my draft notice, we went to the military folks together and suggested that they might be able to use our talents to their advantage. Unbelievably, they agreed, and we were brought in as captains with the sole duty of ferreting out theft occurring here in Vietnam. We've been in-country for nine months now, and this is the first assignment we've been given that won't result in someone being sent off to LBJ (Long Binh Jail) for a well-deserved penal R&R. You really should not have told the IG that he is stupid, although we think that he must be if he is willing to waste our time to settle a personal score. You have no idea how much stealing is going on all around us. We get our kicks out of sending the bad guys to jail, and this waste of time means that one more crook will stay off the hook."

These guys were really very nice and, judging from the stories they told, very good at what they did. They wound up spending another three days reconstructing my books. In the end, they added about $5,000 to the net worth of the mess association. I had written off a bunch of obligations when we converted from COLA to a Class 1 mess (details of that conversion will be shared later), and although this write-off had been an effective way of keeping our current operating statements in the black, regulations required that we make these write-downs over a twelve-month period. Although the rules of the game did not permit them to give me a copy of their final report, they did go out of their way to tell me the general contents.

The gist of the report to be submitted to the Commanding General IV

Corps, the Inspector General IV Corps, and copied to Headquarters Saigon, was in effect that these were the finest set of books they had ever come across in all their military investigations and that Lieutenant Dinan should be complimented on the professional quality found in his recordkeeping. Further, it would state that something in excess of $5,000 was to be added to the net worth of the Can Tho Mess Association due to the peculiar requirements of the military system. They would also certify that no financial irregularities were found and that there was no foundation for seeking their investigation in the first place. All in all, it became a feather in my cap and egg on the face of the IG who called in the investigation.

I received a formal written instruction from the Commanding General regarding my need to comply with the findings and to adjust my books accordingly. No other comments or accolades were forthcoming.

TWENTY-ONE

THE NEXT ACTS OF RETRIBUTION

ALTHOUGH I was smugly elated by the outcome of this ambush, I was not, this time, lulled into a sense of complacency. I knew that there must be another avenue of attack available to the enemy, but I lacked the experience that would have alerted me to what it was. I soon found out.

It was a few weeks later at approximately 0830 hours. I was just contemplating the idea of removing myself from my comfortable bed when one of my minions appeared at my bedside in a state of panic. With no preamble, he announced, "Lieutenant Dinan, you better get over to the mess halls right away. There are some really strange things going on."

My first thoughts were extremely X-rated, but I answered, "I'll be there right away. Tell me, has anyone been hurt?"

"No, Sir. Nothing like that, but still, I think that you should come over as soon as possible."

As I was crossing the parking lot heading toward the mess halls, I could clearly see a large white van about which much activity was in play by persons dressed in white coveralls and surgically masked. I had been told that no one had been hurt, but this sure looked like a medical emergency was in progress. Quickening my pace, I was soon confronted at the door leading into the officers mess by an armed guard who demanded, "Who are you, and where do you think you are going?"

Somewhat annoyed I replied, "I am Lieutenant Dinan, I am the mess officer, and I am going into my mess hall."

In that wonderfully dutiful way that is the joy of any soldier of enlisted rank who is given the opportunity to so address a commissioned officer, he replied, "You will not enter this facility. We are conducting a full-scale sanitation evaluation; your presence is strictly forbidden. You are to remove yourself from this area and to stay away until you are instructed otherwise."

"I am what?"

"You heard me, Sir. Move out."

As he was armed and otherwise evidently officially supported, I moved out and went to my office. On entering I felt a sense of doom and gloom in the air and in my staff. That they were worried on my account was an occasion for rejoicing, but as this sentiment was definitely out of place, I withheld my sense of joy and meekly inquired, "Does anyone know what in the world is going on?"

The general silence was deafening and the attendant gloom gave rise to a feeling of genuine alarm. As everyone else found more opportune subjects for their ocular activities, it remained for Sergeant Butler, who had more time in the army than the rest of us had in life, to look me directly in the eyes and give me the unhappy news.

"Lieutenant Dinan, I have seen these things happen before, and I am unhappy to inform you that you are fucked." Now, there's an unwelcome evaluation of a situation.

"Sergeant Butler, can you expand on this?"

"Yes, Sir. What you have is an investigation by a special operations unit that is charged with finding sanitation and other health related deficiencies that may be found in a food operation. My experience is that they always find enough irregularities to put the OIC in serious difficulty."

Anyone who has experienced a Class 1 inspection in which a team comes in with white gloves and flashlights knows that passing such an inspection is improbable. Anyone who has suffered through the OCS experience has intimate firsthand knowledge of this from repeated personal experience. As one of the above mentioned, I felt doomed. In such a situation, the only thing to do is nothing, and I did this with a keen sense of the fact that there was nothing I could do. Was I worried, concerned, or otherwise discombobulated? You bet. I was scared shitless! Something was always findable, and something would be found. What, who, how, whatever—I had no idea. The only apparent certainty was that I would be humiliated and found negligent in my assigned capacity. In this situation, I succumbed to the usual activities of stress and worry. Certainly these activities are noncompensatory and in every way unrewarding, but what other options did I have?

Time moved forward surely yet very slowly.

At about 1030 hours, one of the white-suited wonders appeared before me and proclaimed, "Lieutenant Dinan, we have completed our inspection and you may resume your normal duties."

Bucking my normal inclination to present an above-the-fray demeanor, which would have engendered a response such as *Good for you; now get lost,* and overwhelmed by a combination of fear and curiosity, I asked, "Is there anything that you can tell me about your findings?"

Naturally an informative response would be too civil and too civilian, and so I was officially instructed: "Our report is confidential. It will be submitted to the Commanding General, the IG, and the Chairman of the Can Tho Mess Association Board of Directors. They will determine what information should properly be communicated to you. Again, you may resume your duties."

With this he went out, and I was left feeling rather left out. As soon as I was informed that the van had cleared the compound, I hustled over to the mess halls to find Sergeant Erickson. Not surprisingly, he and his staff were in a hurry-up mode trying to get lunch ready on time. Although he obviously saw me, he ignored my presence; not a good sign to my way of thinking.

I stayed on the sidelines until I figured that things were under control, and then I approached him, "Sergeant Erickson, is there anything you can tell me?" He asked that I give him a little more time, and this I did. As I watched, I noted that he was, as always, earnest, professional, and totally focused on his duties. This was truly a great Mess Sergeant.

When he was satisfied that everything was under control, he came over to me and said, "OK, I can spare a few minutes now."

With no formal greeting, I let fly the most important questions on my mind: "What did they say? Did they find anything really horrible? What did they ask you?"

Sergeant Erickson looked at me serenely and hesitated before responding. When he'd gathered his thoughts he said, "Lieutenant Dinan, I can see that this is a first for you. I've been through these things before. They don't say anything. They don't tell you what they have found, and they don't ask any questions. They submit their report and move on. When the big brass gets the report, that's when the action starts.

"But let me tell you this, this operation is topflight and I don't know of anything, not just horrible, but anything that they could have found. Now, I have a lot of things to do, and I suspect that you have things to do also, so stop worrying and keep working. All this will take care of itself."

I knew that he was correct when he said that this was a first for me, and I had every reason to believe him when he said that he had been there before. I took his advice and kept on working, but try as I could, the "stop worrying" part failed me.

About a week later, I was absolutely stunned when our Chairman of the Board, U.S. Army Corps of Engineers, Colonel Barnes, burst enthusiastically into my office bellowing, "Congratulations, congratulations, congratulations! This is spectacular! It is the best, best, best, and you are the best!"

I stood up immediately and asked, "What's happening? What's going on?"

He came over and pulled up a chair and with great flair plopped down on my desk an important looking official report.

"Here," said he, "is the best goddamned report any mess has ever received. I'm so proud of you that I could just burst."

I was completely taken and transformed by his enthusiasm. Who wouldn't be? I blurted out, "What does it say?"

He said, "You can read it for yourself, but what it says is that our mess is the finest, cleanest, and all-round best operation that they've ever encountered anywhere, anytime. They say that they were not able to find even one infraction or deviation from military standards and that the General and his staff should be highly commended for the leadership they have provided."

I was thrilled and I couldn't keep from asking, "What about me? Was I mentioned?"

It was instructional to learn, "No, your name is not mentioned anywhere. You have to recognize that although you have to do the job, the responsibility is placed high above you. This is a triumph for the General, just as a bad report would have been a black mark on his administration. You can be sure that he will appreciate what you've done for him."

Sometime later, that appreciation was to rear its beautiful head.

A few weeks later, I returned to Hong Kong to finalize the specifics of our purchases for the mess association. Unlike my first trip, where I had utilized my personal R&R allocation, this was an official business venture requiring official orders. These orders were circulated amongst the entire G staff, which of necessity included the IG, Major Nordquest. The lag time between orders and departure gave ample opportunity for my IG buddy to organize a full IG Inspection of my operations, thoughtfully scheduled to commence at the time of my departure.

As with my first trip, this trip was also an unqualified success.

As with my first trip, which had inspired so much hope in Colonel Fletcher, this excursion served to inspire a similar hope in Major Nordquest, which would ensnare me in a web of gamesmanship propelled by the aspirations of that unhappy and unqualified official.

As with my first trip, I returned jubilant and triumphant to the rural capital of Can Tho and was greeted with instructions to immediately contact a colonel. This time it was Colonel Barnes. I readily complied and was notified that a special meeting of the board was to be convened ASAP to deal with a report submitted by the office of the Inspector General. Could there be any greater joy?

Having assembled and dispatched a team of diligents charged with finding fault in any aspect of my operations, the IG was rewarded with a report specifying exactly thirteen points of administrative failure. These thirteen points were forwarded to the board for their action of reprimand against the OIC, aka me.

The special meeting of the board was scheduled, and I was provided with a copy of the thirteen points to allow my preparation of a defense or explanation. I entered into the review process with fear and trepidation. A careful review of these points soon turned my attitude into one of barely restrainable ferociousness and temper. A more trivial and inconsequential listing of mismanagement or failure to comply with regulations was beyond all imagination. Although the singular thinness of the assembled charges precludes recollection of all thirteen, two were so larded that they can never escape my memory.

One of the charges specified that I was in dereliction of duty because I "allowed and condoned" the duties of my cashier to be accomplished by members of my Vietnamese staff when he needed to be relieved for personal reasons. This practice predated my arrival. It must be noted that these ladies, who were gracious enough to accept this responsibility, were not only fluent in English and wizards at currency exchange, but also highly educated. They were shown great deference by all the locals because of their families' social positions in the Can Tho community. The implications of this challenge were a direct insult to the Vietnamese people whose hearts and minds were our ultimate target. In any event, I had gotten to know and to highly respect the ladies in question; any slight as to their integrity, in this case questioning their suitability to handle money, made me furious.

The other charge forever implanted in my memory concerned a check that I had personally approved for cashing. This $500 check had been cashed without my having the casher provide in writing, on the check, the extensive personal identification information, Social Security number, military ID number, et cetera, required by U.S. Army regulations. Unquestionably, in this instance I was in dereliction of duty. Thankfully, however, the evidence of my dereliction, the actual check, was still held under my authority. I had not yet taken it to Saigon for deposit. It was a personal check from none other than the Commanding General of MACV IV, Major General Eckhardt. Strictly adhering to the military regulations, I was required, as the officer approving this check-cashing transaction, to insist that the General provide all stipulated identifying information just as any other soldier. I judged this manifestation of evenhanded treatment for all to lie somewhere between asinine and tawdry. Technically, I was in the wrong. I should have insisted that the General annotate his check with all the required information. But intellectually, I felt entirely justified.

I arrived at the board meeting at the appointed hour. Upon my arrival, in contrast to the usual casual and informal pre-meeting conviviality, the board was pre-seated and very serious in demeanor. This was to be more of an inquisition than an exchange of information. I expect that I was meant to present myself as a penitent. I did not. I was not penitent; I was pissed. This entire inspection, as well as its report, was a conjured and contrived hatchet job instigated by the IG whose stupidity I had had the stupidity to vocally acknowledge. Instead of assuming the expected role of supplicant, I waged war as a soldier. The meeting having been called to order, I attacked.

I stood and announced, "Gentlemen, before you question me on any of these thirteen points, I will individually address each and every one of them for you." With rage-induced adrenalin surging through my entire body, I channeled its energy to specifically focus on each point. I challenged, humiliated, and defeated each point with rhetorical abandon. I invoked vocabulary intended to incite: insipid, ignorant, unwarranted, stupid, arrogant, foolish, dogmatic, insensitive, innocuous. I was a veritable thesaurus of insult and disdain.

My interrogators quickly became inoculated by my venom and joined in my rage. By the end of my soliloquy, I could almost smell their solidarity. I could taste their anger.

One board member summed it up when he blurted out, "This is all bullshit; we're not going to take this crap. Fuck the IG!"

The ensuing chorus of "That's right, it's bullshit, fuck him" warmed my heart. It looked like the chairman, Colonel Barnes, was going to lose all control over the meeting. He did not.

He let everyone get the bile out of their systems, and when they quieted down, he said, "Gentlemen, I believe we have heard everything we need to hear from Lieutenant Dinan, so unless anyone has a specific question for him, I will dismiss Lieutenant Dinan, and we will continue this meeting. Are there any questions gentlemen? No? Well then, Lieutenant Dinan, you are dismissed. We will notify you of what actions we intend to take in response to these reported findings."

The finality of his dismissal left me no other option than to say, "Thank you, Sir," and take my leave.

The meeting went on for another hour or so, and when it was over, Colonel Barnes came to my office and told me, "Lieutenant Dinan, that was a very unusual presentation for you to give under the circumstances. It was, however, provocative, powerful, and on point. The board has determined to challenge each point in this report. I will be personally overseeing the rebuttal that will be submitted to General Eckhardt. Until we receive his response, you will continue to operate the mess association according to the instructions we have given you."

I said, "Thank you, Sir."

He replied, "No, we thank you for the job you have been doing for all of us. Keep up the good work."

I liked this guy. It was good to know that someone had a clue as to just how hard I worked to keep all the various functions under my control running smoothly.

As promised, the board submitted a reply to the IG's report to General Eckhardt challenging each point. In the General's response he wrote, "During my long career, I have never before rejected in total a report submitted to me by my Inspector General. After careful review, I reject in full this report concerning the operation of the Can Tho Mess Association and thank the board for their careful review of each of its thirteen points."

Now, finally, there was ample cause for celebration. As I celebrated, I could not help but reflect on the possibility that the General was in some real way showing his appreciation for the glowing reports submitted by earlier inspectors.

Some days later I ran into Major Nordquest while en route to the officers club. He stopped me and said, "You know, Lieutenant Dinan, there is nothing personal in all this."

"In all what?"

"You know, the inspections, I'm only doing my job," he replied somewhat sheepishly.

I, lying through my teeth, answered: "My presumption is that, as Inspector General, inspections are your job, and the idea that anything personal was involved never occurred to me. Actually, I suspect that I should thank you because, without your inspections, people might only assume that I am doing an excellent job. It is your inspections that have proven the veracity of that assumption. I trust that when you are inspected, the results will show that you are doing your job equally well."

"Well I, I . . ."

As he stumbled to find something intelligent to say, I saluted him and said, "Thank you, Sir."

With my salute duly returned, I left him standing in place and continued on and into the officers club for a well-deserved scotch. I was to have one more, and final, confrontation with that fellow officer.

TWENTY-TWO

INFORMATION VERSUS DIRECTORY ASSISTANCE

THINGS continued at a usual pace. I hesitate to use the word normal, as that would incorrectly imply that the on-the-ground reality of our lives at Eakin Compound was similar to that of our civilian lives. I continued to make the military mistakes usual to usual second lieutenants. Vietnam at that time had three telephone companies that did not work well together. It could take hours to actually cause the phone to ring at the office or on the desk of the person one needed to contact. To mitigate this inconvenience, senior staff officers were assigned callers whose sole duty was to call, call, and recall until the desired person could be found on the other end of the line.

Totally frustrated by the telephone system but far too junior to be provided with such a caller, I resorted to telephoning the Information Officer for directory assistance.

When I reached the Information Officer, I was gratified to hear the respondent declare, "Office of Information, Lieutenant Coupe speaking, Sir." As I proceeded to enumerate the telephone related information I needed, I was taken aback by the seeming deadness on the other end of the line.

When I took a pause, I was greeted by this inquiry, "May I ask who is calling?"

I immediately responded, "Lieutenant Coupe, this is Lieutenant Dinan from the Can Tho Mess Association."

In reward, I received a rather caustic, "I see. Tell me, Lieutenant Dinan, is your branch of service the U.S. Army?"

"Yes it is."

"I am Naval Lieutenant Jay Coupe, and I outrank you by far more than you could possibly understand, and I do not give telephone information. I provide information to the press, to Congress, and to other important parties. Don't you dare ever call me again with your stupid questions."

Bang went the phone, end of call. And I thought, silly me, I just wanted a little help. I had been brought up with the understanding that when you needed help finding telephone numbers, you called directory assistance. If you were from New York, you called Information. It was so easy to be so wrong.

Not long after that, I found myself enjoying a drink in the officers club. It was one of those evenings when the club was particularly crowded and convivial. Into this boisterous enclave stepped an individual garbed in Japanesque regalia complete with elevated footgear suitable to a grade C movie. Given little notice in spite of his garish attire, this person moved to the middle of the room and began to sing. At first no one noticed. Then this one, then that one, and another one stopped their talk and listened. This unusual phenomenon was singing "Che Gelida Manina," the great tenor aria from the first act of Puccini's opera, *La Bohème.* During the elongated musical pause in this aria, the only sound you could hear was that certain intake of breath that screams, "Please don't stop," and at the precise, exact, correct musical moment, this a cappella triumph resumed. At the end, the eruption of appreciation was almost as spellbinding as the performance itself. Taking a deep bow and turning to his left, the performer attempted a speedy retreat. This was not to happen. The exit was blocked by an appreciative and motivated audience unwilling to release this giver until more was given.

The peculiarly attired performer, having extracted promises of release following one more offering, relented to the crowd and offered up a rendition of the classic Irish heartbreaker "Danny Boy" in such a way as to rend the heart of any listener. When he pealed into the second stanza and admonished us all:

> And if you come, when all the flowers are dying
> And I am dead, as dead I well may be
> You'll come and find the place where I am lying
> And kneel and say an "Ave" there for me,

wet eyes were rampant, and many a hardened warrior reflexively made the sign of the cross without thought of external observation. In that interlude of overwhelming silence, our performer disappeared into the night.

My first thought was *Who was that masked man?* I asked around, but it seemed that no one had the least idea. Someone had to know, and I was determined to find out.

So I asked and asked until I was rewarded with this intelligence, "Oh, that's Lieutenant, excuse me, Naval Lieutenant Jay Coupe, our new Information Officer, freshly delivered from Naples, Italy."

My heart sank. I should mention that I am a big opera fan with a particular warm spot for tenors. I wanted to make this guy a friend, and I had already gotten off on the wrong foot with our unfortunate first contact via the telephone.

I waited several days, trusting that I would run into him casually, either in the mess hall or the O club. This did not happen because, as I was to find out, Lieutenant Coupe reported directly to the General, and he was part of that exalted group referred to as the generals mess. Being so privileged has its poisons. The poison attendant to this privilege is the requirement that one spend the bulk of his personal time in attendance with the General and his staff. Escaping from the exalted bonds or bands of this camaraderie with any regularity was considered a direct affront to the General and carried a heavy penalty. It's just one of those things that nobody ever even considers doing.

Thus it was that finding a one-on-one situation that would include my target would require a little imagination. I was determined.

I called the Information Office and once again I was greeted with, "Office of Information, Lieutenant Coupe speaking, Sir."

I announced myself, "This is Lieutenant Dinan calling, Sir." When he replied, it was difficult to reconcile the ugliness of his tone with the beauty of that voice that still rang in my ears.

"Lieutenant Dinan, I believe that we may have had this conversation before. I instructed you then and I will instruct you again, this one last time. I do not provide telephone numbers and you are not to call this number. Do you understand me?"

I felt like this might be my last chance so I charged ahead, "I need important information, not a phone number."

With smug surprise he countered, "And what important information could that possibly be?"

It was clearly my moment and with some trepidation I ventured, "What must one do to learn to sing half as well as you do?"

"Oh?" said he. "I'm not sure that I can answer that question, but I thank you for the compliment."

This had a positive ring to it so I forged ahead, "Listen, I understand that you are tied up virtually all the time, but my quarters are air-conditioned, we have a full bar, a great sound system, and I'd like an opportunity to get together whenever it's convenient for you."

"I'll tell you what. Give me your phone number and I'll see if I can get back to you." I gave it, and just prior to hanging up, he said, "Thank you, and, Lieutenant Dinan, remember not to call this number ever again." Promptly the line went dead, and this ball was taken well off my court.

Some days passed, and I had just about given up hope when, answering my phone, "Can Tho Mess Association, Lieutenant Dinan speaking, Sir," I was rewarded with the response, "Lieutenant Dinan, this is Lieutenant Coupe, and unless it is inconvenient, I will present myself at your quarters at

2100 hours this evening." It sounded great to me, and I said so. It seems that when the number 21 appears in my life, it foretells glad tidings.

At the appointed hour, my guest arrived, and he allowed as he was suitably impressed by the cool comforts peculiar to this domicile. I, in turn, allowed that although he was the best living tenor I had ever heard in the flesh, Jussi Bjoerling was my very favorite of all time. He concurred and from this fragile link a strong friendship developed. Without any formal acknowledgement, the lieutenant thing was dropped. He was Jay, and I was Terry. Although his bondage prevented our getting together with great regularity, he was to become a relatively frequent and always welcome guest. Jay was one of those people who can best be described as a Star Performer in life. That said, a little of his bio is necessary.

Jay Coupe had indeed been delivered to us directly from Naples, Italy. He regaled me with stories about his favorite watering hole in that great city, Da Umberto, and I was inclined to suppose that he exaggerated the establishment's participation in that love affair. This was in 1967, and when Anne and I found ourselves in Naples in 1970 we made a point of going to dinner at Da Umberto. Taking a chance, I asked our waiter if he might remember an American naval officer named Jay Coupe. Did he remember? You'd think I'd spoken of the Second Coming. Everything in the place went on hold. The entire staff gathered around us as if we were emissaries of some great potentate. Most of the ensuing exclamations were in Italian, so I can't vouch for exactly what was said. The facial expressions and the body language did the real communicating, and the repeated phrase "*Che bella voce!*" had serious impact. It appears that my friend had indeed been modest.

Jay went on to have a highly distinguished military career, retiring with the rank of naval captain. If you see some of the footage showing our prisoners of war being escorted out of North Vietnam in 1973, the fellow escorting Senator John McCain—that's Jay. When he was later stationed at Governor's Island in New York, he frequently visited The "21" Club, where I had joined the management, and our home. In his final assignment, he was the senior aide-de-camp to Admiral William J. Crowe, Jr., who was then the chairman of the Joint Chiefs of Staff. So highly did the Admiral regard this officer that, when Jay was leaving the service and entering into marriage, he hosted a huge celebration in his palatial personal residence quarters at Fort Myer, Virginia. Anne and I were happy attendees at that very special reception and were privileged to view the Admiral's personal collection of ceremonial military headgear that spanned both the centuries and the globe. This was a great retiring act for a Star Performer.

TWENTY-THREE

THE MESS CONVERTS AND MY PROMISED ASSISTANT ARRIVES

BETWEEN arias and air raids, our war continued its advance. At a certain point, we in MACV IV, Team #96, Eakin Compound were judged to be so very advanced in carrying out our roles as modern day, state-of-the-art warriors in the Delta, that it was determined that we should be supplied with food in the same manner as most of the other U.S. military forces serving in Vietnam. This would require conversion from a COLA system to a Class 1 mess. As the command concerning this changeover was made by the top honchos in Saigon, there was no room for argument. You are ordered to convert, and you will convert.

All conversions are fraught with many difficult lifestyle decisions. We had a pretty good system going in which most of us found comfort and solace. The mess halls, or dining rooms, were operated essentially as adjuncts of the clubs. It was certainly unusual to find a military feeding facility that used plates rather than mess trays. It was equally unusual to find waitresses instead of wait lines. A wine list and a charcoal broiled steak, cooked to order, have never been the hallmarks of a traditional military feeding facility. Class 1 did not in any way signify first class. Were we, subjected to an enforced conversion, to abandon our first-class accommodations and humbly submit to a Class 1 prescribed style of dining? Such an appalling possibility was completely rejected by this mess officer.

Working together for our mutual benefit, the board and I devised a system for allowing revenues from the clubs and the billeting services to be used to support our civilian staff in the dining facilities. The mixing of funds is well outside usual practice, and to satisfy various objections, the system was altered and tweaked over time. Although we were close to being forced to operate as typical mess facilities, during my term of service we always maintained full service dining rooms.

In making the changeover to a Class 1 mess, Sergeant Erickson and I faced many daunting challenges. In order to sustain the level of menu diversity and choices that we were accustomed to presenting in our dining rooms, a substantial backup of supplies had to be maintained. According to regulations, a mess hall was not allowed to have a warehouse where large quantities of supplies could be maintained. A mess hall was authorized to have a small storeroom for dry goods like salt, pepper, sugar, and other incidental items. Our a la carte dining system required a decidedly larger storage facility. To overcome this challenge, we resorted to time-honored tradition and opted for the most sensible course of action—we cheated. We false-fronted our largest storage facility as a carpentry shop and packed the place to the rafters with everything we could lay our hands on. In time this became an open secret, and when the Class 1 facility at Can Tho Airfield was in dire straits, we were able to lend them supplies so that no one in the Delta went hungry.

A brief description of how the Class 1 supply system worked is probably necessary. A number of mess hall operations are assigned to a central distribution facility. In this instance, we, along with a number of other operations, were assigned to draw our rations from the distribution center at Can Tho Airfield. A ration is one meal: a breakfast, a lunch, or a dinner. If I serve 100 of any of these meals today, tomorrow I get to requisition 100 replacement meals. If 150 hungry soldiers show up, I am to make do and the next day I can draw 150 rations. What exactly I get depends upon the menu for that day as prescribed in the universal ration guide. One of the little catches in this system is that if a particular item specified in the guide is not available, you can demand a substitute that will match the computed dollar value of the missing item or items. As the system functions, you have to be ready to fight and know what to fight for, or you will wind up with nothing or with something useless to your needs. Successful ration procurement is not to be entrusted to those short on valor or imagination. Under Sergeant Erickson's sterling leadership, our procurers had no equals.

Another challenge that confronted us was what to do about the teams in the field that had come to depend upon us as a major resource for their food-buying needs. I made the decision to keep giving our support to these comrades who lived in the greatest danger with the least of comforts. I was given the go-ahead to continue giving this support with strict orders to ensure that no commingling of product would ever be tolerated. We had a storage building near the checkpoint at the entrance to the compound. This we expanded, augmenting it with CONEX refrigerator and freezer units, as well as display shelving and other accoutrements needed for a small grocery store.

While all of this change was in progress, my long-awaited assistant/second in command arrived in the person of Second Lieutenant James King. I was thrilled at the prospect of having some much-needed assistance, and once again I was impressed by Colonel Bagley, who had kept his word, even though he was no longer assigned in Can Tho. My joy was soon shattered when the powers-that-were informed me that they had reconsidered the most productive use of Lieutenant King's services. It was determined that a separate dining facility would be created for the General and his staff and that Lieutenant King would be in charge of that facility. This generals mess, which would service approximately a dozen people out of the more than three hundred regularly assigned to Eakin Compound, would require an officer dedicated solely to that purpose, so Lieutenant King would not, in fact, be available to assist me. Adding frosting to this cake, in the special operation orders written for Lieutenant King, it was specified that if he should encounter difficulties requiring assistance he should first come to me. I was charged with the responsibility of addressing and fixing any and all of those problems. Happily,

Jim was a really good and capable officer and rarely needed my help. On the other hand, I would have loved to have had him helping me. Alas, the military moves in mysterious ways.

TWENTY-FOUR

THE GROCERY STORE

I must mention that the go-ahead I received regarding my support operation for the five- and six-man teams in the field did not include any official recognition. There was no budget, no chain of command, no authorization for personnel, and no authorization for transport. The only official cover we had was the aforementioned statement in the rules governing mess associations to the effect that one mess association may sell goods to another mess association in an emergency situation and that the seller is allowed to charge 5 percent over his basic cost to cover the various expenses usual to the procurement process. We tallied all purchases at our original cost and to the bottom line we clearly added 5 percent. This modest surcharge covered our expenses and at the same time saved the supported teams considerably more than they would have spent going to Saigon for supplies. So successful was this operation that we were at one time giving support to as many as 108 teams. In fact, word of this operation filtered up to the highest levels in Saigon, and a serious effort to emulate its success was initiated.

One day, two lieutenant colonels presented themselves at my desk and announced, "We are from the supply and support command in Saigon, and we are charged with duplicating the operation you have here for sub-sector support throughout the country."

I was astounded, and as my chest puffed up, I stood and said, "Gentlemen, how can I help you?"

I was taken aback by the response, "You can conduct us to the officer-in-charge."

I said, "I am that officer." At this juncture I was the recipient of two of the most incredulous stares on record. One of the gentlemen, in a voice pungent with disbelief, managed to utter, "But you're only a lieutenant."

I acknowledged, "That is my rank, Sir."

With a withering glance I was told, "We do not seek advice from someone of your rank."

With this, the two turned on their heels and left. I was informed severally that the numerous operations opened to emulate the service we provided failed miserably. The military moves in mysterious ways.

The amount of actual work and administrative effort that went into sustaining this support operation was staggering, but we all felt compensated by the reality that we were giving our help to the most endangered of our comrades.

The MACV IV Command had divided the IV Corps area—the Mekong Delta—into geographical sectors. In turn, each sector was divided into sub-sectors. The overall command of the sectors was headed by either a Special Forces command or a MACV command. Within these command structures, the sub-sector teams were often mixed, with Special Forces units reporting to MACV or vice versa. Over time, the complaints from MACV units under Special Forces control became so regular and consistent regarding the poor, or lack of, support they received, that the issue had to be adjudicated by the highest command in Saigon. We were all dumbfounded when word came down that the decision on this matter was in effect that Special Forces were not required to provide any support to any unit that was not a Special Forces unit. This unbiased ruling also noted that MACV was charged with giving full support to any Special Forces personnel operating within the theater of their command. Go figure.

Chiefly because the number of MACV sub-sector units vastly exceeded the number of Special Forces units operating in the Delta, the bulk of our clients were MACV. At the same time that the MACV units were lodging complaints to their leaders because of lack of support from Special Forces, my subordinates were lodging complaints to me about our Special Forces clients because of their thievery.

I knew, as did everyone, that these folks were commonly referred to as The Chicken Thieves in recognition of their much heralded acumen in procuring by stealth and skullduggery rather than cash. The PX Officer, Lieutenant Carlson, who had been one of my roommates, had often told me of his storage facility at Can Tho Airfield being broken into and looted in the darkness of night. Each time, it was determined that it had been a Special Forces operation and that therefore no action would be taken because this sort of extracurricular activity was an important ingredient in the overall goal of keeping their wits sharpened. This continued to be confirmed by Lieutenant Myer, his replacement both at the PX and in C-5. So I counseled and consoled my staff, advising them to heighten their alert when visited by these very special fellow Americans.

My guys were good, but the competition was better trained and more motivated. We were regularly looted by these special people who were actually applauded for the skill with which they executed their appalling behavior. I'll say that we did cut it down, but we could not cut it out. There was some grumbling, and I was questioned as to why we should work so hard to help people so dedicated to stealing from us. The only answer I could conjure was the unreasonable assertion that they were *special.* An untenable hostility was brewing, and some sort of showdown or confrontation was unavoidable.

One sultry afternoon, two of my guys returned from Saigon on a commandeered C-130 aircraft loaded with supplies destined for our grocery. Once again we begged and borrowed a two-and-a-half-ton truck and met them on the tarmac at Can Tho Airfield. Sultry is too polite a word to describe the blistering heat and humidity that greeted us as we alighted from our borrowed truck. We were further greeted with the miserable news that the airfield forklift was out of order; we would have to unload this mountain of goods by hand. Considering the overall horror of the situation, I abandoned usual military protocol and physically joined my enlisted men in the labor of unloading the aircraft and loading the truck.

We hadn't gotten too far along when a jeep occupied by two Special Forces sergeants, nattily resplendent in starched fatigues bearing the weight of many up and down stripes on the arms, requisite Silver Wings Upon the Chest, dashing aviator sun glasses, and topped off with green berets set at exactly the most jaunty angle, came screeching to a halt just beside us.

Disdaining the execution of a salute that military courtesy requires of an enlisted man approaching a commissioned officer, the passenger, while remaining in his seat, announced, "It's too bad the forklift isn't working."

I readily agreed, and he continued, "We have our own forklift and it's working just fine."

"That's great; bring it over," I said. "Unloading by hand in this heat is pure torture."

To my astonishment, he replied, "It will cost you a couple of cases of food."

Completely taken aback, I asked, "Sergeant, do you know who I am?"

"Sure, you're Lieutenant Dinan, I know exactly who you are."

Now this special genius was starting to really get to me, so I inquired, "Do I then assume that you are aware of the number of Special Forces teams that we supply? Do you realize that these supplies are intended for that purpose?"

"Look," he said, "I already told you that I know exactly who you are, so of course I know what this plane load is for. It's very simple. If you want to use our forklift, it will cost you a couple of cases of food."

It was hot, and at this point I was hotter. The thought that went through my mind began with *You surly son of a bitch.* Happily I was able to restrain myself, and instead I said, "Sergeant, you have acknowledged that you know who I am and that my operation gives a great deal of support to a significant number of Special Forces units. Further, you have acknowledged that you recognize that these supplies are intended for that purpose. Do you agree with what I have just said?"

With great self-assurance he replied, "That's a Roger."

"Then Roger this, sergeant. Either you bring your forklift over here and we use it with no payment, or when you return to your unit, you tell them that Lieutenant Dinan will no longer sell to Special Forces personnel."

Through the dark glasses I could not see the look in his eyes, but the curl of his lips was even more expressive than his words: "Talk's cheap. You can't do that, and unless you're willing to pay, we're not willing to play."

Barely containing myself, I dismissed him with the curtest of military dismissals, "That will be all, sergeant."

I turned my back, and they screeched off back whence they had come.

My boys were thrilled. They all asked if I really meant what I said and if I could get away with carrying through on my threat. I assured them that I meant every word, and that there was no way I would continue to break my balls for that group. There was general rejoicing, and I was told that they would prefer to unload by hand every time if it meant that they would not have to service those beanie-wearing thieves and blowhards.

This was a particularly large consignment, and the unloading job under the circumstances was punishing. Notwithstanding the sweat and exhaustion, spirits continued to climb with the expectation of freedom from serving this group. It was only at this juncture that I became fully aware of the enormous hostility and resentment that the proactive procurement procedures peculiar to Special Forces units had inflicted upon my staff. The more I learned of their intimidating and intransigent behavior when dealing with my staff, the more committed I became to my decision to no longer deal with this *select* segment of our military community.

Over the next few days, my guys were particularly joyful in reporting their satisfaction when refusing service to several Special Forces teams seeking to make purchases. I suffered no illusion that this action would be somehow be spared the requisite equal and opposite reaction. Some forceful challenge was inevitable, and a forceful challenge was launched.

Mid-morning, while I hunched over my desk toiling to reconcile various ledgers, a phalanx of heavily armed men, signaling their unflinching unity by dint of their headgear, the unparalleled Green Beret, stormed into my office and established a perimeter about my desk. This silver-winged scouting party, comprised of a full half-dozen plus one stalwarts, presented noteworthy diversity. Here we had not only one major, two captains, and one lieutenant, but also representatives of the enlisted ranks, two sergeants first class, and a staff sergeant. Once the line was formed, the major, having assumed the center position, advanced a step forward and addressed me.

Utilizing a familiar military intimidation technique, he opened his challenge with the words, "Soldier, you have some answering to do."

I had been expecting a visitation although I had not envisioned one of this menacing magnitude. Giving due military respect to the object before me, I stood and declared, "Lieutenant Dinan, Sir. What question would you like me to answer?" This was undoubtedly an unexpected response, and I could sense that the major was somewhat unhinged, if only for a moment.

Recovering admirably he responded, "You will tell me how you dare to presume to tell my people that you will not provide them with your support."

I replied, "Sir, there is no presumption here. Your people, as you refer to them, have regularly harassed my people by stealing anything they could lay their hands on while making purchases at our facility. Although we have tolerated this intolerable behavior beyond toleration, there is now the added fact of the outrageous and unprecedented rudeness shown to this officer by your senior enlisted men that has caused us to sever our relations. I informed your people as to the consequences of their insistence that I give them cases of food in exchange for the use of your forklift at Can Tho Airfield, and they all but spat in my face. Special Forces are very special, and you can take special care of yourselves, Sir. I, for one, am no longer engaged, Sir."

The major was enraged, "My men have only acted in accordance with the way they have been trained. You have no authority to do what you're doing; you will sell to my people, and you will do so immediately."

I suppose that I was simply too angry to be intimidated, even when confronted by so much superior rank and firepower, so I simply stated, "Sir, my sales activity is not mandated; it is a courtesy that requires a great deal of effort and work. I am no longer willing to subject my people to giving their effort and energy to a pack of discourteous thieves. Your people will have to find someone else to harass with their procurement techniques. We will no longer sell to you."

The major was aghast. His cohorts looked to me, to him, and back. Taking another threatening step forward, he presented me with a face contorted in violence and demanded, "Soldier, is that your final say?"

"Sir, I have nothing more to say, Sir."

"Well I do, soldier, and you have not heard the last of this matter."

Turning smartly on his heal he marched out, followed by his bewildered entourage. My office staff, U.S. and Vietnamese alike, was all atwitter, and I was left shaking like a leaf. I knew that more would come of this, and I knew that it would be very unpleasant. Nonetheless, on reflection, I felt rather satisfied with the confrontation and began to prepare mentally for what was surely to come.

When answering my phone a short while later, I was hardly surprised to hear: "This is the First Sergeant speaking. You are ordered to report to the compound commander immediately."

Immediately I reported, and immediately I was received, only to find Major Magnus appearing disoriented and nervous as usual.

He looked at me as if I were an alien from outer space and said, "I don't know what's going on, but it seems that you have outdone yourself. Colonel Hazen is enraged about something you have done, and he has ordered me to personally escort you to his office."

We proceeded directly to the compound commander's jeep, and I deposited myself in the back while he took the front right-hand seat next to his driver. As we sped out of Eakin Compound en route to Command Headquarters, I could sense by the particular gravity with which the Major gnawed on his warts that an unpleasant confrontation was in the offing. This was to be my first meeting with the new Deputy Commander, Colonel Hazen, aka Shazam, and it was to prove turbulent.

Upon arrival we were immediately ushered into Colonel Hazen's office, previously the lair assigned to Colonel Fletcher. Not much had been altered, yet one could not help sensing a totally different atmosphere. Shazam had a reputation as a down-and-dirty military man of the old school in contrast to the effete and elaborately sanitized Colonel Fletcher. The essential difference hit me in the face like a mud pie. This was the office of a real military man, cut from the same piece of cloth as those celebrated in our favorite movies. This was the real thing.

Allowing for no nonsense, Colonel Hazen immediately instructed me to be seated. When I went down, he came up and launched an unnerving verbal attack: "What is this total bullshit I hear about you disrespecting and refusing instructions given by a senior military officer?"

This neatly worded question forced me to defend myself simultaneously on two fronts. The act of disrespecting a senior officer was considered so egregious an affront in military life as to be beyond forgiveness. Refusing instructions as opposed to refusing a direct order was somewhat less cut-and-dried. Glaring rather than looking at me, the Colonel was clearly requiring a prompt and satisfactory reply on both fronts. The atmosphere was highly charged; I was well aware that this was a serious situation, which should not in any way be taken lightly.

And yet there was something unrepentant lodged within my sense of humor that compelled me to observe, "Sir, I sit here accused of two serious offenses."

My use of *sit here,* rather than the standard *stand here,* was not at all lost on Colonel Hazen. He offered me a rather bemused look and then proceeded to survey his surroundings. I followed his eyes and we both alighted upon Major Magnus at the same moment. The rumpled Major was huddled against a wall studiously ignoring our interchange while seriously engaged in gnawing upon his knuckles. Following his surveillance, Colonel Hazen seated himself, and with a slight twinkle in his formally ferocious eyes, looked me in the eye and leveled me with this observation, "You have indeed been accused of two very serious offenses, and this may be your only opportunity to defend yourself. What have you to say in your defense, Lieutenant Dinan?"

"Sir, although it hasn't been stated, I think I am correct in assuming that this meeting is connected to my meeting earlier today with Special Forces personnel. I assure you that at no time did I show any disrespect to any person in that group that invaded my office. If there is any allegation of disrespect, I think that we should address that issue first."

I was very much disarmed when Colonel Hazen responded, "Let's put that horseshit aside for the moment and concentrate on the real issue. How in this world do you come up with the idea that you have the authority to decide who can and cannot buy goods at the Eakin Compound commissary?"

"Sir," I said, "there is no such thing as a commissary at Eakin Compound."

This declaration seemed to confuse the Colonel. He leaned way back in his very plush chair, and for several moments closed his eyes and clasped his hands behind his head.

Shortly, his hands came down smartly upon his desk, and half-rising from his seat, he leaned toward me with flashing eyes and demanded, "Then tell me Lieutenant Dinan, what exactly, the fuck, do you have going on there?"

I was stunned by this invective and for some moments I could author no response.

The Colonel's patience was short-wired, and while I was still in the recovery mode he launched another fusillade: "Don't sit there looking stupidly at me. I'm waiting for your answer."

Regaining my wits, I answered, "Sir, we operate a support activity that is intended to be helpful to the many small teams we have out in the field. It is unofficial and has no specific authorization. Right now we support over eighty teams, and they really appreciate our efforts. They tell us that we save them a great deal of time and money, as well as helping them to eat much better than they could without our support. Keeping this operation up and going requires a lot of work, but we all feel really good about what we're doing."

As he took an extended pause, I came to realize that, amazing as it seemed, he really had no idea what our not-so-little grocery operation was all about. Taken aback by his own lack of knowledge, he returned to the subject at hand stating, "It appears that you are very proud of this unofficial operation that you have been carrying on."

Although this was clearly an observation that begged for a response, it was not a definitive question. My military intelligence training had instilled in me the precept that you never answer to a question that has not specifically been asked. But in this case I felt compelled to provide a response.

"Sir, we are all very proud of what we have been doing."

His hands returned to the back-of-the-neck position, and he turned his eyes toward Major Magnus, possibly seeking some clarification. I followed his glance and discovered that the Major remained huddled against the wall, his eyes averted, and his attention clearly focused upon his unfortunate feasting activity. Colonel Hazen wasted little time in this venture, and in a moment we found ourselves locked, eye-to-eye. This was in no way a stare-down situation. I was being assessed, period. Colonel Hazen had that unique look-deep-into-you ability that I had experienced with Colonel Bagley. My God, how we all hate that feeling of being completely naked in front of a fully clothed superior officer.

He further contributed to my discomfort by asking, "Let's suppose for a moment that this unofficial operation that you have been conducting is not something that will land you in the brig, and mind you, I will personally have that aspect fully investigated. Where do you find the nerve to tell our Special Forces operatives that *you* will not support them?"

The hostility directed at my nerve or presumption was laid bare on the table, but as I was overheated, I was in no way intimidated by the threatened jail time. My ire enjoyed an all-time high. I was inflamed. I gave the Colonel, chapter and verse, all the indignities we had suffered at these special hands including the unrelenting thievery and the shakedown at Can Tho Airfield that had sealed the deal. He pondered my soliloquy, and I could sense that he recognized my pride of accomplishment along with my frustrations. However, he was a big fish and as such, he was frying fish larger than my frustrations.

Giving ceremonial acknowledgement to my grievances, he specified, "We have a big war going on out here, and what you have been up to seems to be a good thing. Unless I determine it to be otherwise, I'll let you continue to do what you have been doing. In the meantime, you will reconsider your obstinate and unacceptable position regarding who will and will not have the benefit of your support."

Clearly, I was being instructed in no uncertain language. Equally clear was the certainty that I was committed to giving no support whatsoever to our Special Forces brethren.

I found myself saying, "Sir, I will not continue to break my balls to do anything for these thieving beanie-sporting troops."

Hearing this noble declaration, Major Magnus tore himself from his feasting with such force as to gain our attentions. He did not say anything, but everything about his body language warned me to say yes to anything and no to nothing.

The Colonel seized upon this moment to seal his purpose: "Lieutenant Dinan, I don't give a rat's ass about the supposed indignities you are convinced that you have suffered. You like to think that the support that you're giving to our troops in the field allows you to puff up your little chest by thinking that you are in some way contributing to the success of our overall mission. At the same time you have the arrogance to suppose that you can elect who you will and will not support. Listen here, and listen good, you will support *everyone* or you will support *no one.* Have I made myself clear to you, Lieutenant Dinan?"

"Yes, Sir. Thank you, Sir. You have, Sir."

With this yet-to-be-acted-upon understanding, we swiftly departed, and chauffeured by his driver, Major Magnus and I wound our way back to Eakin Compound.

Although I wanted to be dropped off at the grocery, Major Magnus insisted that I accompany him to his office for a debriefing about our meeting.

After gently suggesting that I had to be completely out of my mind to utter any words other than "Yes" and "Sir" when a full colonel presented me with his recommended course of action, he went on to inquire, "Well, Lieutenant Dinan, just exactly what do you plan to do now?"

"Sir," I replied, "when you dismiss me from this meeting, I will walk over to the grocery and put a lock on the door. As of immediately, I plan to avail myself of the option to serve no one."

His disappointment was palpable and he asked, "Do you think this is a smart military decision?"

I acknowledged, "I suspect that it is probably a stupid decision in military terms, but as a human being, it is the only thing I can do."

He warned, "I think you're being driven by a hotheaded and unwarranted passion. I strongly recommend that you give yourself time and reconsider what you are about to do. Remember this will be your decision, and you will stand alone in suffering the consequences."

I thanked him for his advice, and he sent me on my way. My way was directly to the grocery. I felt that the Major was completely correct on the point that I was possessed of a hothead. On the other point, the one about my passion, I disagreed entirely. I knew that my passion was inflamed, and I felt as though my entire body might explode. Yet I was certain that my passion was well founded, and I was determined to govern myself accordingly.

There was electricity in the air when I arrived at the grocery. It was that very recognizable, that intangible, all-pervasive aura of expectancy written plainly and unwillingly upon the afflicted who sense that some momentous action is imminent. Pretending that they were completely unaware of what was going on, my men offered me a lighthearted welcome, "Hey, Lieutenant Dinan, what's up?"

"What's up is that this operation is going down. Give me your keys and the lock. Remove your paperwork to the main office. Take the rest of the day off and report to me at 1000 hours tomorrow."

Betraying their pretense of ignorance, no questions were asked. I received lock, keys, salutes, curious looks, and no questions. When my staff departed I took a careful look about, turned out the lights, carefully closed and secured the outer door, and then, with unpremeditated violence, slammed the lock into place. I marched myself back to my office, deposited the keys in my desk drawer, and retired to my rooms, availing myself of the comfort of air-conditioning to cool my body and the comfort of a long cool drink to cool my mind.

TWENTY-FIVE

THE GROCERY SAGA CONTINUES

WHEN my grocery workers duly reported to me at 1000 hours the next day, I assigned them to work with Sergeant Cole, who was the NCOIC of club operations. I knew that Sergeant Cole was stretched at all times by the enormity of his duties and that he could always use an extra hand. I also had the sense that this would be a temporary assignment because I believed that the substantial benefits afforded by my grocery store would preclude a permanent shutdown. I simply knew that something big time would have to happen. I had no idea as to what that would be.

Over the next several days, I was interrupted regularly by those sent on shopping missions from various sub-sector teams. Overwhelmingly my interrupters were senior NCOs from the various advisory teams, as food shopping was considered to be an enlisted man's arena unsuitable for direct regular physical involvement by the elite commissioned officer ranks. The basic question put to me by these professional soldiers, although delivered in diverse terms, had a universal consistency that added up to "What the hell is going on?" In each instance, I took the time and gave full chapter and verse in explanation of the current unhappy situation. The reactions were consistent. There was no antipathy, but indeed there was anger. The universal consensus could be summed up as, "We have once again been screwed by the Special Forces," and the universal declaration insisted, "When we report this to our officers, all hell is going to break loose."

I was really touched by the reactions of these disappointed professional soldiers. To a man, they reaffirmed my belief that my men and I had been doing something meaningful and important to the success of their missions and for their comfort. Further, they reaffirmed the righteousness of the negative feelings inflicted upon me by the unseemly activities of this thieving band. I knew that something would happen; I just did not know what that happening would be.

Several days later, upon returning to my office, well-refreshed, having lunched and "tennised" (an invented word—it had become my habit as well to engage in two sets of tennis during the two-hour midday interlude), I received another summons from Major Magnus. Once again, he was to escort me to the office of Colonel Hazen. When a full colonel summons a lowly lieutenant into his presence under escort, the prospects of a cordial encounter are nonexistent. Owing to this certain reality, I accompanied the Major filled with fear and trepidation. Once again, upon our arrival we were ushered immediately into the Colonel's private office. I sensed that our ushers, although acting with the highest level of military courtesy and aplomb, were equally in a state of high angst bordering on fear. Distinctly pressured by this highly charged atmosphere, I entered into the lair in a spirit of desperation.

The Colonel, though seated, exhibited an aura of such physical activity as to further alarm my entire being. I, we, were not offered the courtesy of a seat. We stood, helpless as two ants in front of a behemoth, suffering a look of disdain so powerful that any combative spirit I possessed was wilted entirely. Had I been asked to speak, I don't suppose that I could have summoned up one single word.

As in the former meeting, the Major's presence was ignored. Vaulting from his seat and extending his entire person so aggressively and violently across his desk, all the while affixing his blazing glare so heatedly into my eyes as to cause me to shudder in fear, he assaulted me with this momentous declaration, "I don't know what the fuck is going on here."

My peculiar sense of humor kicked in and I immediately thought, *What else is new?* Something related to my prior training or current fear caused me to stand at even more rigid attention and remain silent. When it occurred to me that my eyes had become so very glued upon the Colonel's burning orbs that a stare-down contest in which I could not compete was in the offing, I lowered my eyes and broke off the hostile contact. In recognition of this victory, the Colonel resumed his seat.

I looked about for Major Magnus, hoping I might gain some support or guidance. My search revealed that the Major had taken a step and a half back and was engaged in his favorite pastime of wart noshing. I was very much alone.

Thus I was standing alone and at attention in front of Colonel Hazen, who, now comfortably seated, took what seemed to be an inordinate amount of time staring at me inquisitively. He had those eyes, those eyes, the ones that bear down into your soul and prevent you from telling an untruth. I felt, once again, that sensation of being naked and uncomfortable. I just stood there feeling humiliated and exposed. What an unimaginable power is somehow acquired or invested in our senior military officers.

After allowing himself what must have been the exact prescribed amount of time needed to assess my person, position, attitude, or whatever combination thereof, the Colonel said, "I can see that you have no idea what's been happening here."

Somewhat stunned, I said nothing and gave him what I suspect was a hopeful look. He gave me that curious and inquisitive look and asked directly, "What exactly did you say to the MACV teams when they asked why you closed down the store? Think carefully now, and don't even think about bullshitting me; I want a straight answer."

I took a very deep breath and then proceeded, "Sir, when I was asked why I closed down the grocery, I explained that when I excluded the Special Forces teams because of their thievery and attempt at extortion, you gave me the option of either welcoming them back as customers or of serving no customers at all, and that I had chosen to serve no one rather than service the Special Forces."

He was quiet, very quiet, and it seemed like an eternity as he simply looked at me, presenting a stony countenance that gave not the slightest hint of any emotion. This was my first encounter with a true poker face. It was frightening, like looking into a death mask. Nothing moved; no sign or signal suggested that behind that face there remained a living, feeling person. I could feel my knees start to weaken. The sweat ran down from my armpits like they were open faucets sprung from nowhere. I was sweating, and he let me sweat. What he was doing, by doing nothing in the extreme, was cruel. I couldn't help but think, *this is a hardened combat soldier; he's killed before and he will kill again.*

Just when I had reached that point where I was about to break down completely and scream, *Say something, anything,* Colonel Hazen interrupted my misery by saying, "Very well, Lieutenant Dinan." In the long pause that followed, the only thing I could think of was, *Very well, what?* He gave me a smile that was somehow other than warm and commented, "For a United States Army Officer, you are a strange duck. We both might find out just how strange as time goes by. As of right now, the most important thing on my mind is the support of my troops. You will leave my office, and you will go back to Eakin Compound, and you will open your goddamned grocery, and you will give your support to every MACV team who shows up at your doorstep. That's an order, and if I have to revisit this subject one more time, you will be the unhappiest son of a bitch in the entire U.S. military. Have I made myself clear to you, Lieutenant Dinan?"

"Yes, Sir. Thank you, Sir." I presented him with a smart salute.

Rather than return it he just stared at me and said, "You may go now." I went.

I bolted out into the sunshine with such single-mindedness that I completely forgot about Major Magnus. As I turned about to find him, we collided, and he said nothing other than, "Get in the jeep." As we sped back to Eakin Compound the Major asked, "Should I drop you off at the grocery?"

"No, thank you, Sir. I need to get back to my office to get the keys and round up my men."

It was evident that the Major wanted nothing more to do with this entire affair because he asked nothing further and separated himself from me as quickly and as decisively as possible.

I had noticed that there were quite a few jeeps surrounding the entrance to the grocery as we passed on our way into the compound. When I got to the office, I was surprised to find the interior and exterior littered with majors and captains. I pretty much recognized most of them because I had met with them when they signed guarantees for the purchases to be made for their teams. I don't know who was protecting the field because most of the field power was assembled right there.

I was assaulted immediately with the redundant demand, "What's going on?"

I reveled in my response, "The store is officially reopened as of right now."

We opened up the store posthaste, and the ensuing shopping frenzy was a joy to behold. In the midst of it all, I became acquainted with the cause of Colonel Hazen's conundrum. In response to the store closing, sixty-odd team leaders, all majors or captains, had held a powwow that resulted in the decision to approach the Colonel en masse to confront this grievance. Upon his arrival at his office that morning, the Colonel had been alarmed to find the bulk of his field team leaders assembled in military formation in front of his office. A spokes-team had been elected, and that angry group had persuaded Colonel Hazen to grant them an immediate interview. The negotiations held in his private office were reported to be, as the saying goes, down and dirty. The spectacle created by the formal formation maintained by the balance of the powwow participants outside inspired unrelenting calls from not only all the G offices, but even the General himself, who had a call placed to find out what was happening. This was not a happy morning in the life of Colonel Hazen. By the time the negotiating party departed, an agreement had been reached that the Special Forces could look after themselves and that the Colonel would order me to reopen the store for the benefit of the MACV teams. I never discovered why I was subjected to my horrid meeting, and I never again sold anything to anyone topped off by a green beret.

TWENTY-SIX

THE GOOD LIFE GOES ON

LIFE in the Delta was not entirely circumscribed by that ring of turmoil that the Germans describe as Sturm und Drang—storm and stress. Following in the firm footsteps of those who went before us, we found ways of inserting many elements of civilization and joy into our drab green army lives.

Great and unexpected help in this challenge was provided by Rich Andresen, my closest boyhood friend from Evander Childs High School in the Bronx. Rich was an electronics geek, a ham radio operator who could also hot-wire the family car in ten seconds. Later in life, Lieutenant Andresen, United States Army Signal Corps, found himself in charge of a signal company situated along the DMZ between North and South Korea. Finding written communication with yours truly to be a hopeless endeavor, this now Captain Emeritus from Delta Airlines, and still close friend, took charge of the situation and sent to me in the Mekong Delta, a voice tape recorder. Those little voice tape machines seem positively Neanderthal today, but at the time they constituted the height of sophistication. Intrigued and delighted by the possibilities offered by this device, I quickly secured a similar recorder at the PX and sent it along to the red-headed beauty who is now my wife (married six months after I returned stateside).

At the time, my regular residence was C-5, Eakin Compound, Can Tho, Vietnam, and Anne's regular residence was House One, Sweet Briar College, Sweet Briar, Virginia, U.S.A. Quickly our tape-recorded communications became communal activities, and the correspondence developed into social events for the occupants of both of these diverse domiciles. The arrival of a tape recorded in a far-off civilization by those Sweet Briar sweethearts was cause for much celebration in C-5. Giving proper respect to this benefit, we would first refresh ourselves at the officers club and then proceed to dine in the splendor of our officers mess, selecting the finest of wines from our limited combat-zone–fine-wine collection to further lubricate our respective speaking apparatuses. Thus enhanced, we would retire to our luxuriously air-conditioned quarters. First we would assemble an adequate backup supply of liquid ingredients necessary to concoct further tokens of civility. Then we would settle back and enjoy all the commentary provided by those very special gals who were ensconced in the bucolic beauty of the Blue Ridge Mountains. Our personal enhancement was generally so well accomplished that we undoubtedly made very little, if any, sense at all when we recorded our replies, comments, and questions. But what fun! How special! What camaraderie! These were very special events for us and a valued reminder of just exactly what was so very special about the U.S. of A.

During my military sojourn as an enlisted man, where I had succeeded in gaining the rank of corporal in the Intelligence Corps, I was stationed at Fort Holabird, Maryland for a year. In many ways, it was a great experience and tour of duty. Besides learning all that cool intelligence stuff, we Private E-nothings were given the benefit of twenty-four-hour Class 1 passes that allowed us to leave post at any time day or night, provided we were not specifically assigned to report for some educational or work-related activity. Our regular weapon while on post was an attaché case, and we were instructed to don civilian garb when venturing outside of the post facility. Going about the city of Baltimore disguised as civilians was intended to camouflage our military association. This spook mentality lent credence to the idea that "military intelligence" was an oxymoron, inasmuch as our required military haircuts easily distinguished us from the local populace, especially in the mid-1960s.

I came to invest some of my freedom and free time in the honorable sport of fencing. I had done some fencing in college, so when I discovered that the local YMCA was offering lessons at affordable rates, I signed up. The fencing master was a gentleman of Hungarian origin and a boyhood friend of Giorgio Santelli. Giorgio, born in Budapest in 1897, was the son of Italo Santelli, and Italo was reputed to be one of the world's greatest fencing masters ever. This master, Stephen Bujnovsky, studied alongside his boyhood buddy Giorgio under his father Italo, and at the time I undertook these lessons, he was also the assistant fencing coach at the U.S. Naval Academy in Annapolis. Steve was gloriously Hungarian. He presented himself in striking black fencing attire, and forgoing any attempt to waste his valuable time learning the names of his supplicants, he addressed all his students as "Daaarling." His rules were quite simple. You did exactly as he instructed or you went somewhere else. When you went on the mat with the master, you did not wear protective head gear unless so instructed. To insist otherwise was an insult, and insults were not tolerated.

The approximate fifteen minutes one spent on the mat with the master seemed a lifetime, at the end of which it was difficult to determine whether it was one's mind or body that was the most exhausted. Suffice it to say that when dismissed, you could hardly stand, and you could not begin to think. After a few months of instruction, I was informed that the master would no longer work with me unless I would commit to two lessons per week. Given my meager circumstances, I was reluctant to reach further into my impoverished pocket, but the master was relentless. He insisted that he had taken me as far as he could within the limits of once-a-week instruction and that he would not waste his time with me unless I was willing to commit to a twice-weekly regimen. I reconsidered and I relented, and so it began in earnest.

"Daaarling, when I do this, you will do that, and this, that and this, that and this/that and that/this. Now again, Daaarling, this/that, this/that, that/this, that/this. Again Daaarling, again Daaarling, again Daaarling." One day, I was shocked to be told, "Daaarling, you will put on your mask."

It was tantamount to being told that you could now wear long pants instead of the short pants reserved for little boys. Wow! Now I am a grownup!

It became incredibly more relentless: "This/that, that/this, that/this, this/that, Daaarling, Daaarling, Daaarling, again Daaarling."

And then it happened, from out of nowhere my world went into slow motion. It's almost impossible to describe, but I saw every move of his perfectly placed weapon as if it were happening very slowly. My mind comprehended, and I was able to make a judged decision as to my proper reaction and execute the proper move with what seemed like time to spare. Previously, it had been all reflex and hope. This was a coldly calculated and mind-driven response.

The master recognized the difference immediately and exclaimed, "Daaarling, now you are truly fencing. Again, Daaarling, this/that, that/this, this/that, ad infinitum."

I was ecstatic. I had heard about athletes being "in the zone" and I thought it was all nonsense. The master had brought me into the zone, and it remains an unforgettable experience.

As mentioned earlier, Eakin Compound was replete with athletic paraphernalia, including an athletic director. Noting a lack thereof, I appealed to the director to provide our post with fencing equipment. This appeal was unceremoniously denied as a poor use of limited funds. To resolve this dire conflict between my desires and the budgetary constraints imposed by my government, I personally arranged and underwrote the procurement and shipping of a full range of equipment necessary for the pursuit of the ancient and honorable art of fencing.

At that moment in history, there was no paucity of military officers who had been instructed in the art of swordsmanship, and I had the joy of finding many willing opponents. To record the outcome of my engagements might be considered an exercise in braggadocio. Therefore, off the record only, I can happily report that I prevailed at every encounter. These exploits in ancient military madness did not go unnoticed. Pictures and commentary of these engagements found their way into at least one of the newspapers produced by and for the American military personnel serving in Vietnam: *THE DELTA ADVISOR,* October 1968, VOL. 1, NO. 6.

My considered view is that the double achievement of introducing fencing along with Baked Alaska did, in fact, enhance our cultural contributions to our allies in the Delta. It should be further noted that our Vietnamese allies concurred with the greater majority of their American military advisers in the consensus that this particular lieutenant was, in a word, different. This, of course, was to me unquestionably the most positive of all accolades.

As we plodded along doing this and that for the benefit of our fellow Americans, the "advised"—the Vietnamese—determined to open a particularly spectacular club for the benefit of their own officer ranks. At some point, the Vietnamese military hierarchy in Can Tho came to realize that in conjunction with the grand opening of their facility, they could in no way execute the production of the food and beverage services in such a manner as to garner accolades from the intended beneficiaries. Thus challenged, this knowledgeable hierarchy appealed to the ever-present and uniquely situated Mr. Hein. At the same time, others of the highest rank in the Vietnamese Armed Forces made their appeal for assistance to their corresponding military compatriots who naturally included my commanding general. The search for remedial measures that might be quickly taken pinpointed me as the most logical source of assistance. Given the honor of being called upon to provide support to the most important of our advised comrades in arms by such powerful enthusiasts, I was unable to find an honorable way of declining the mission.

Since this was definitely something I did not wish to do but rather had to do, I determined to inject an element of sophistication into the financial aspect of the implementation process. The exacting support requirements—menu, beverages, and staffing—were very carefully worked out and agreed upon. The next hurdle was in the final detail—payment. How much? Who? In what way? These are the usual considerations in any catering negotiation.

How much was fairly negotiated and fairly easily agreed upon. *Who* was designated as the Vietnamese High Command, worked out in such a manner as to preclude possible further inquiry as to any specific individual who might assume that responsibility. *In what way* proved to be more contentious. There were two aspects to consider: *how much upfront* and, because of my determination regarding the financial part of the process, *which currency* was to be used in payment. The upfront figure I suggested caused the usual bitching and moaning and resulted in the usual contraction of the requested amount until an agreement was reached. The currency to be used was a special case.

As mentioned earlier, there were two legitimate operating currencies circulating in South Vietnam. The Vietnamese piaster was the currency to be utilized for all indigenous transactions. The military payment certificate was the authorized currency for all U.S. government personnel to be utilized for all transactions at U.S. military installations. While the Americans and affiliated service personnel of other allied countries were entitled to the possession and use of these MPC, Vietnamese of any rank, stripe, or character were not. Because the MPC was connected to the U.S. currency and the piaster was connected to the "kleptocrats" in Saigon, these MPC became looked upon as more secure paper holdings than the local currency. The Vietnamese desire for MPC was even further increased because, with very little effort, those of significant position could easily cajole an American associate into making purchases on their behalf at the nearest PX, as long as the cajoler had the required MPC. The cost and availability of such items requisite for human dignity such as cigarettes, alcohol, toilet paper, shampoo, and other basic necessities, served to fuel a frenzy of accumulation of this U.S. special-issue currency within the highest ranks of the Vietnamese military establishment.

Everyone knew that the Vietnamese military hierarchy was hoarding MPC, but it was inconvenient to officially acknowledge this. When I suggested that payment should be made in MPC, everyone involved went berserk. When questioned by my superiors, I explained that this presented the perfect opportunity for us to rectify, or at least make a dent in, this illegal and potentially embarrassing currency problem. As it was the military's duty to police the use and distribution of our currency, they unofficially agreed, but officially they maintained the position that their counterparts were to be considered innocent of such skullduggery. In the end, they succeeded in pretending that it was an issue outside of their control. The problem was clearly Lieutenant Dinan's, and the Vietnamese negotiators would simply have to deal with me.

The Vietnamese went directly to Mr. Hein, and I suspect that they twisted his arms in every possible direction.

Mr. Hein came to me and asked, "How can you ask these officers to pay in MPC when you know that they are not supposed to have any?"

He must have been put under a lot of pressure because he even went so far as to provide the *oriental face* story and noted as how I would be causing a great loss of that commodity by blatantly demonstrating that I assumed that they were guilty of this serious infraction.

"Look," I told him, "you and I both know that these guys are holding tons of MPC, and they know we know. So instead of all this hand wringing and whining, why don't they just cough it up?"

We went back and forth, Mr. Hein and I, and in the end, I was paid 80 percent in MPC and 20 percent in piasters. The best was yet to come, and the groaning was to be replaced by rejoicing.

It was about two or three weeks after the grand opening of the new club and the settling of the final bill that the great powers in Saigon sprung an MPC conversion on the entire country. Evidently, the sanctity of the MPC had become so breeched that it had caused major concern for those charged with control of such things. To remedy this terrible inconvenience, the color of the currency was changed, and a very rigid exchange process was instituted. The time frame for exchange was very limited, as was the roster of those who could make an exchange. The intended result, which was to put the proud but unauthorized hoarders of MPC in the unenviable position of holding worthless paper, was realized. In the ensuing period, there was great crying and gnashing of teeth. After all, loss of face is one thing, but loss of cold hard cash is the real thing.

I became a hero to the local Vietnamese High Command. Mr. Hein thanked me profusely on behalf of the senior Vietnamese officers, who refused to be convinced that I had no prior knowledge of the monetary Armageddon. No matter how many times or how many ways I asserted my complete ignorance, the reaction remained: "Of course, we know. We understand. You would never do anything like that, and we thank you very much."

The rejoicing and joy expressed was somewhat colored by the regret that I had not been tougher and stuck to my guns and insisted on a 100 percent payment in MPC. Eventually, they apparently concluded that they had been just a little too tough for me. The important thing was that they had been spared the loss that the 80 percent MPC payment represented.

At least once a month, Mr. Hein would insist upon taking me out for dinner. On one occasion he even entertained me at his home. This was quite a spectacular event, what with all the staff and family, and course upon course of I have no idea what, only that it looked and tasted wonderful. Generally, we went to Chinese restaurants, of which there was a great abundance. I would be picked up by Mr. Hein in his chauffeur-driven, air-conditioned, brand-new Volvo and whisked out of Eakin Compound in great style and comfort. Our custom was to make a stop to pick up his good friend Major Doe, and the three of us would descend upon some chosen eatery. Wherever we went, Mr. Hein was given the VIP treatment, and we would invariably be seated in a special room reserved for the celebrity class. These rooms would be in the back somewhere and well protected with sandbags and other anti-villain fortifications. I never had a meal that I didn't like, and I never was able to eat as much as my host or his friend. Although neither man weighed in at more than ninety pounds, they could literally eat me under the table.

On one occasion Mr. Hein insisted that we go to *the* Vietnamese restaurant. According to my host, actual Vietnamese restaurants were few and far between. This particular establishment was reported to be in a class with Michelin-rated–three-star houses. This was where the most famous of Vietnamese movie stars and other notables would congregate. I'm not sure exactly what I was expecting, but I can assure you that it was other.

The most salient aspect of the décor was that, once seated, you began to notice that the walls and ceiling appeared to be in constant motion. I was to learn that this motion was the result of hundreds of little lizards, which gave me a bit of a start. Evidently, maintaining this brood took a great deal of effort, but the result was that there were no flying creatures whatsoever. Should some reckless insects be bold enough to enter these hallowed grounds, the lizards would make short work of them. The place was very open, there were no screens, and with the help of a few elegant ceiling fans, the air was reasonably comfortable.

All great restaurants have a house specialty, and this one was no exception. The specialty of this house was turtle. These turtles were carefully selected and then starved for at least a week in order to insure that the inner tracts were cleaned out. At precisely the right moment, they were steamed alive until cooked to perfection. Although I do not recall what the other notable and highly recommended special offerings were, I do recall being altogether stricken by the mere thought of dining upon those various exotic creatures. I was graciously let off the hook on all accounts, save that of the turtle. My Vietnamese escorts insisted that I was to partake of this most special of specials. We had many, many courses, and certainly more food than I could possibly cram into my mouth. As usual, I was bested in this manly sport by my two ninety-pound comrades. I must admit that everything was delicious.

When the time had come for the most important dish of the meal, the pièce de résistance, our server, a young, barefoot Vietnamese lad clad in shirt and shorts, with great pride and fanfare presented the turtle. Our server proceeded to place the fully intact creature in the center of the table, thereby allowing for the requisite oohs, aahs, and other expressions of rapture. When the ceremony of rapture was done, the shell of the beast, about eight inches in length and of suitable width, was unceremoniously ripped away and slung on the floor beneath our table. Due to miracle or practice, this shell bypassed all shins. The body was expertly sectioned and laid out on a service platter along with all the inner organs and stuff. I was instantly offered the intestines which, like all edibles that appear particularly revolting, are guaranteed to richly enhance one's manhood. At age twenty-three, I was typically over enhanced and I deferred to Major Doe, who wolfed them down with great relish. I settled for what I suppose you might call a drumstick. I found it to be far too gelatinous and gamey for my taste, but it was okay, and I was well congratulated for my gustative adventurousness.

On one occasion, our entire office staff booked a party room at one of Can Tho's finest hotels. The nature of the event escapes me, but for some reason, I was the guest of honor. We were all seated at one large oval table overlaid with crisp linen and adorned with beautiful flowers. As with all Vietnamese feasts, this was a many-course event. The memorable course for me was the poultry course. Entire roasted birds—in this instance, chickens—were presented carved and arranged on serving platters, then laid before us on the table. Entire in this case included the head on its long, distinctive neck.

The young lady in her smart *ao dai* seated to my left quickly snatched up the aforementioned morsel with her chop sticks and demurely placed it squarely upon my plate explaining, "This is reserved for the guest of honor."

This guest of honor, although suitably honored, was unsuitably horrified by the prospect of dining upon this thing.

I therefore, with determination rather than grace, managed to get hold of the beast with my very own chop sticks and delivered it onto her plate explaining, "In America, it is only the ladies who eat head."

I remain uncertain whether or not the obvious double entendre made any impression. What is certain is that the young lady had no intention of eating this thing either and, having exclaimed the requisite "*Choi oi!*" returned it with great speed and dexterity to the serving platter. It's hard to be completely certain, but judging by the tone of the exchanged remarks, which I could not decipher, and the tone of the giggles, which is universal, I suspect that I was the brunt of the age-old game: Let's see what our visitor from another world will do when graciously set up to eat something gross.

Ribbon cutting at the newly refurbished Eakin Compound Officers Club

From left: General Eckhardt, Colonel Barnes, Sergeant Major Marcille

From left: Sergeant Cole, Lieutenant King, Colonel Hill, Lieutenant Dinan

Sergeant Butler, successor to Sergeant Sulivan

Sergeant Chavez

"Chief" Graham, successor to Chief Warrant Officer Jay Arnold

Sergeant Chavez, left, with his replacement, Sergeant Erickson, right

Naval Lieutenant Coupe, center, singing with Lieutenant Dinan
and the Eakin Compound Chaplain, 1967.

John S. McCain III is escorted by Lt. Cmdr. Jay Coupe Jr., public relations officer,
March 14, 1973, to Hanoi's Gia Lam Airport. (AP Photo/Horst Fass)

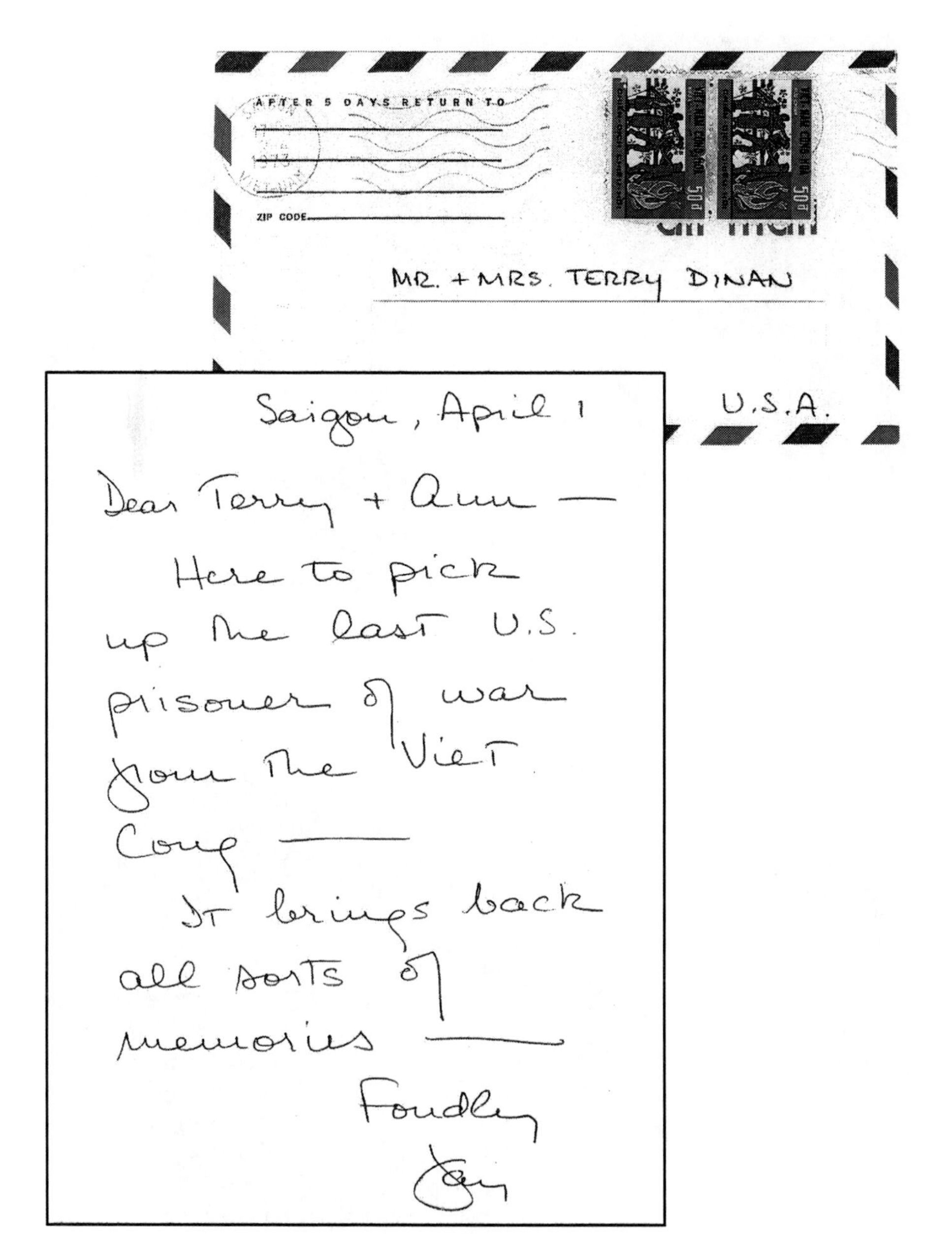
AFTER 5 DAYS RETURN TO

ZIP CODE

MR. + MRS. TERRY DINAN

U.S.A.

Saigon, April 1

Dear Terry + Ann —

Here to pick up the last U.S. prisoner of war from the Viet Cong —

It brings back all sorts of memories —

Fondly

Jay

Jay made several trips to escort U.S. prisoners of war home from Hanoi.
Here is his April 1 letter, postmarked Saigon, 1973.

Lieutenant Myer and the PX staff

Lieutenant Carlson, Myer's predecessor at the PX

Anne and roommates recording a tape for C-5

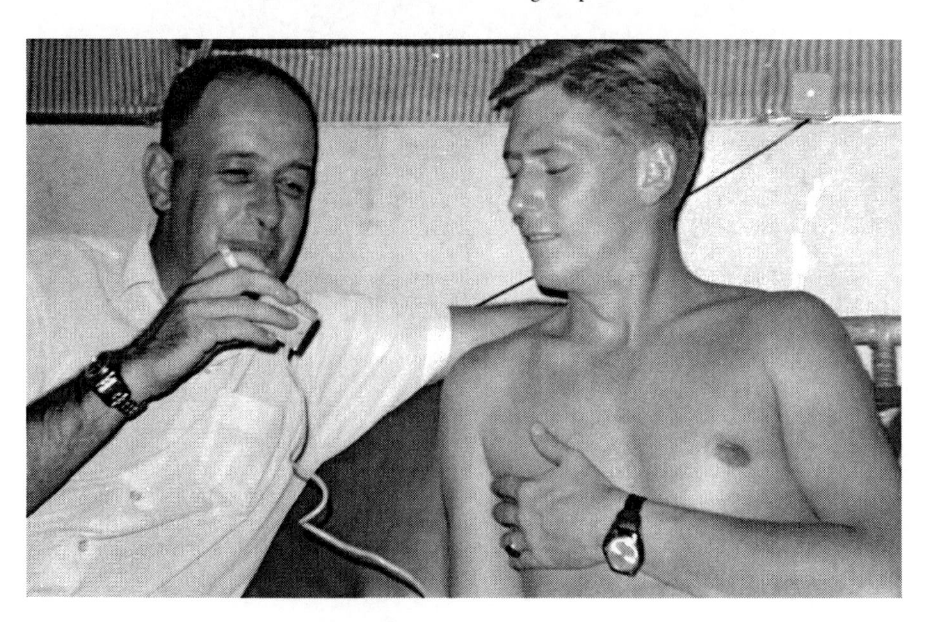

C-5 bunkmates, Captain Dery and Lieutenant Dinan,
recording a tape for House One

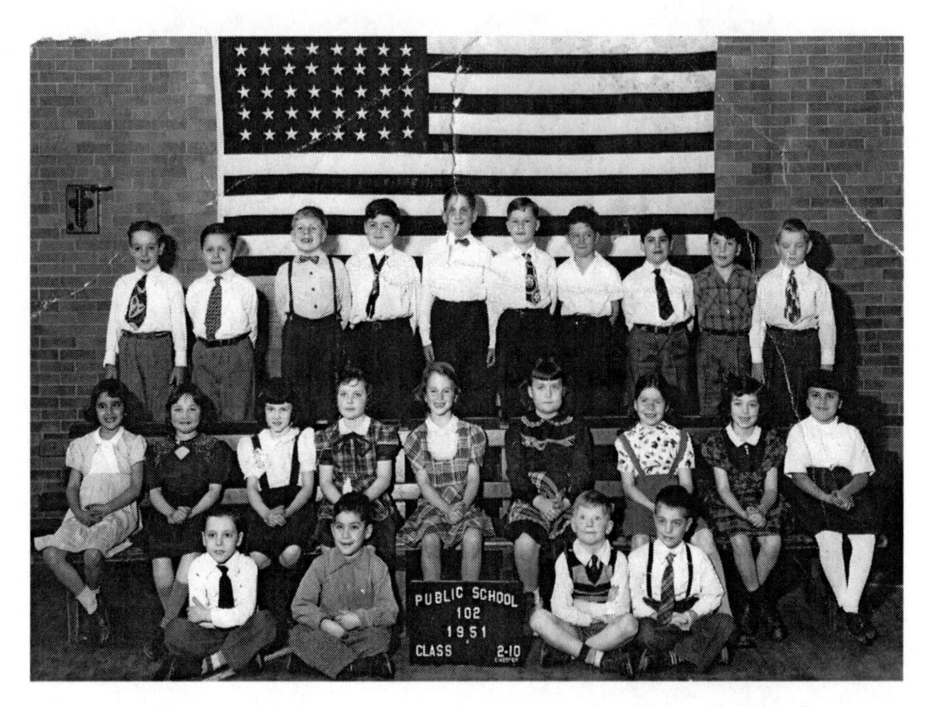

The allusion to short pants is true: Terry Dinan, age 7, first row, is the only boy in short pants, his mother having decreed that he was still too young to wear long pants.

Major Devlin, left, and Lieutenant Dinan

Parry

Score

31st Regiment Commander Honored

THE DELTA ADVISOR

VOL. 1, NO. 6 — Can Tho, Vietnam — October, 1968

31st

LTC Hung Gets Second Silver Star

VI THANH — Lieutenant Colonel Le Van Hung, commanding officer of the 31st Regiment, 21st ARVN Division, was awarded the first oak leaf cluster to the Silver Star Medal in ceremonies in Vi Thanh, Chuong Thien Province.

Colonel Gene A. Walters, senior advisor to the 21st, presented Colonel Hung with the award for gallantry in action during the Tet offensive in February.

In Soc Trang

During that period Colonel Hung, newly designated as commander of all Vietnamese military forces in Ba Xuyen Province, was assigned the mission of clearing and securing the city of Soc Trang and surrounding areas which were heavily occupied by Viet Cong.

The award cites that "throughout the period, Colonel Hung moved about with no thought for his own safety, to effectively deploy his men and encourage them to keep continuous pressure on the enemy forces. At one time, while observing a group of his troops pinned down by machine gun fire, Colonel Hung exposed himself to the hostile fire and knocked out the enemy position.

"Colonel Hung's brave action, personal example, and sincere devotion to duty contributed enormously to a successful operation and resulted in very limited friendly casualties and equipment losses."

A native of the Hoc Mon District, Gia Dinh Province, Colonel Hung, 35, has been in the Army for more than 14 years. He already possesses 16 Crosses of Gallantry as well as numerous other awards, including a first Silver Star and the National Order of Vietnam, fourth class.

New Market Place Opens in Phong Phu

PHONG PHU—A ribbon cutting ceremony officially opened a new market place in Thoi Thanh Village (town of 0 Mon) in this Phong Dinh Province district.

Construction of the market place was a major self-help project with assistance provided by the U.S. Navy Seabees. Major Stanley L. King is district senior advisor.

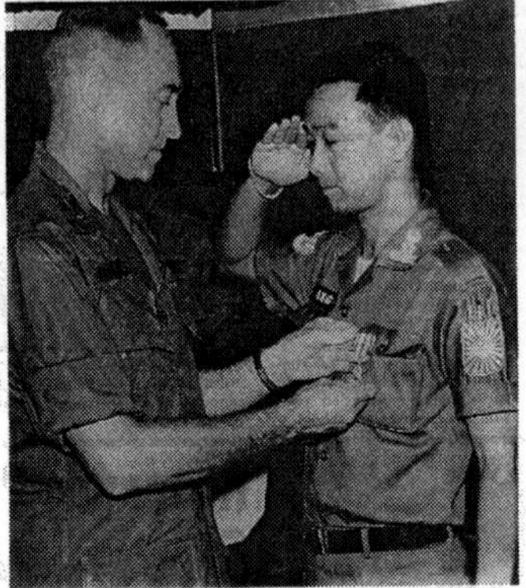

PRESENTATION—First Oak Leaf cluster to the Silver Star is awarded to Lieutenant Colonel Le Van Hung, commanding officer of the 31st Regiment, 21st ARVN Division. Colonel Gene A. Walters, division senior advisor, officiates at the ceremony held in Vi Thanh, Chuong Thien Province. (Staff Photo by TSgt J. Summers)

Fete Children

Festival Big Success

Autumn Festival—also referred to as Children's Tet—proved a success in the Delta earlier this month with a helping hand provided by wives of top corps officers, U.S. advisors and their counterparts and U.S. units.

Mrs. Nguyen Thi Bach Tuyet, wife of the corps commander, hosted a festival party on her front lawn, presided over several others and flew to the "boondocks" to deliver gift packages.

Phong Dinh Province advisors had a massive campaign to collect clothing, toys, candy and money and 4th Area Logistics Command advisors held a Turkey Shoot to raise funds.

Also conducting drives to give children a happy holiday were the 164th Combat Aviation Group, 69th Engineer Battalion, Binh Thuy Air Force Base, Naval Support Activity at Binh Thuy and the 4th Riverine Area advisory team.

In mid-September the campaign was started in Phong Dinh. Major Decatur W. Morse of Jacksonville, Fla., S-5 advisor, was campaign coordinator.

Major Morse placed donation boxes in dining halls, clubs, theaters and other key places throughout the Can Tho military complex. He contacted public service officials at the larger installations to get their cooperation in the campaign.

The donation boxes were emptied regularly. They yielded bags of candy, some clothing, a few toys and MPC and piaster donations.

A small mountain of clothing was provided by the Catholic Relief Service. Fourteen machine-compressed bales of clothing weigh-

FESTIVAL PAGE 8

RF-PRU Force Frees POW's

Regional Force troopers and an An Xuyen Provincial Reconnaissance Unit (PRU) combined to attack a Viet Cong Prison Camp to highlight action in the IV Corps area during the past month.

The force was heli-lifted into an area six miles north of Song Ong Doc in An Xuyen Province to free 22 prisoners including one women, 16 other civilians and five Viet Cong. The force also captured one female VC and three suspects.

Three Years

The civilians had been accused by the VC of being government spies and had been utilized to tend the VC rice paddies. Some had been prisoners for up to three years.

Corps officials termed the operation highly successful and praise the RF and PRU troopers who had never operated with the use of helicopters.

Meanwhile other government forces involved in Operation Quyet Chien were successful in killing VC where the enemy was contacted.

A series of clashes in Sa Dec Province in one day resulted in a total of 62 VC killed. The 9th ARVN Division soldiers detained four others and captured three crew-served and 15 individual weapons and a large quantity of ammo, mines and documents.

Kill 39 VC

The division's 14th Regiment, supported by APC's, had a good afternoon in Vinh Binh Province killing 39 enemy in a fight south of Tieu Can District.

Again, 9th Division troopers scored, killing 31 VC in an engagement near Ba Cang District in Vinh Long Province. Regional Force, reconnaissance and APC units combined with the ARVN soldiers in the battle.

There were many other operations—of all sizes involving various government troops—which left from 10 to 26 VC dead.

A 9th Division operation in Vinh Binh killed 17; a 7th Division clash in Dinh Tuong killed an additional 17; 12 more died in a Ranger operation in Kien Phong; RF soldiers killed 16 in Vinh Long; 26 were counted by 9th Division soldiers in Sa Dec and on the same day in Kien Giang 21st Division troops killed 18 and located an arms cache.

Locate Cache

Soldiers from the 21st Division's 31st Regiment uncovered a significant arms and munitions cache four miles north of Kien Long District, Chuong Thien Province.

The cache included mortars, automatic weapons, 39 individual weapons, 78 cases of assorted ammunition and a large quantity of rocket and recoilless rifle rounds.

Viet Cong incidents continued at a moderate pace with observation posts and watchtowers the prime targets.

PRISONERS PAGE 8

RD MEDALS—Vietnam President Nguyen Van Thieu presents RD Medals to Revolutionary Development cadre during a visit to Ba Tri District, Kien Hoa Province. The president also presented arms to a civil defense group and presided over the formal opening of RD programs.

15-Man Force Defeats Foe

VINH BINH—A self-defense force succeeded in defending its hamlet from a Viet Cong attack earlier this month.

An estimated enemy platoon attacked Ta Cu Hamlet during the early morning hours utilizing AK 47's, automatic weapons and grenades.

The 15-man defense force drove off the attackers and blood stains indicated enemy casualties. Two women in the hamlet were killed by grenades.

TOUCHE—Various sports are undertaken by residents of Eakin Compound to stay in good physical shape. These two officers were spotted one noontime utilizing the volleyball court to fence. Major Edward T. Devlin, left, G-4 plans and projects advisor, duels with First Lieutenant Terrance R. Dinan, custodian of the Can Tho Open Mess. Other sports popular at Eakin include basketball, handball, tennis, volleyball, archery, swimming and weight lifting. (Staff Photo)

In Singles

Buencanino Takes Eakin Net Event

CAN THO—A civilian employee from the Philippine Islands walked away with the Eakin Compound singles tennis championship here.

Vicente F. Buencanino won three out of four sets in a scheduled five-set match with Lieutenant Commander Michael W. Chapple of Coronado, Calif.

Buencanino took the first two sets in the match, 8-6 and 6-3. Chapple rebounded to take the third, 6-2, then dropped the fourth, 6-4.

Chapple posted a 6-1 record in regular tournament play. "My only loss was to my nemesis, Buencanino," commented the Navy officer.

Buencanino was never in the losers bracket of the double elimination tourney. He entered the finals with a 5-0 record.

Net action in Can Tho didn't stop with the singles finals. Play in the Eakin Compound Doubles Tennis Tournament, sponsored by IV Corps Special Services, got underway a week before the end of singles competition.

The tournament opened with 16 teams entered. Midway through the second round one team has been eliminated and another has dropped out.

The following is a listing of the doubles teams and their won-loss records:

TEAM	WON	LOST
Collins-Wiener	2	0
O'Neill-Buch	2	0
Eckhardt-Riess	1	0
Horton-Fouts	1	0
Buxbaum-Weatherstone	1	0
Young-Thomas	1*	0
Chapple-Cornell	1	1
Hughes-Buencanino	1	1
Thompson-Pearson	1	1
Burrell-Semrad	0	1
McDade-McDonough	0	1
Venden-Carlson	0	1
Wright-Salonske	0	1*
Pullin-Sickler	0	1
Quick-Verlautz	0	2†

* Won by forfeit † Eliminated

FINALISTS—Eakin Compound singles tennis champ Vicente F. Buencanino, right, is congratulated by Lieutenant Commander Michael W. Chapple, runnerup in the singles competition. (Staff Photo by TSgt J. Summers)

Delta Advisor Sports

Captain Ragland Discusses Sports

Open Tennis Event Good Idea

BY CPT RAY DERY

CAN THO — A onetime basketball and tennis star in West Virginia believes that allowing amateurs to play against professionals in a national net tourney was a "good idea."

Captain Ned H. Ragland of Beckley, W. Va., and a member of the corps G-4 advisory staff, once played in the Junior National Amateur Tennis Tournament where he faced Chuck McKinley, now a pro, and Don Dell, captain the of the U.S. Davis Cup team. He didn't last long enough in the tourney to meet Rod Laver.

He won the Mid-Atlantic Junior Doubles championship back in 1956.

This year's edition of the national tennis tournament in New York was open to both professionals and amateurs and was billed as the U.S. National Open Tournament. Ironically, an amateur, Army Lieutenant Arthur Ashe, captured the title.

"Having amateurs meet professionals might induce more potentially good American players to stay with the game as a profession," Captain Ragland explains. "Why not let a man capitalize on his skill," he asks.

The captain, who rotates this month, believes that Ashe will some day become a professional.

An all-state hoopsters at Woodrow Wilson High School in Beckley, Captain Ragland played both basketball and tennis as a freshman. Two sports were too much so he selected tennis playing on the varsity for three years.

As a frosh hoopster he played alongside Rod Thorn who is now in the National Basketball Association. The captain, at 6 feet, was one of the shorter men on that squad.

In talking about basketball, Captain Ragland feels the American Basketball League's three-point field goal is good and says the NBA should adopt a similar rule.

Every time a high-scoring team is held down by a deliberate five, fans and sports writers start discussing methods of speeding up play.

Captain Ragland commented on possibly adopting something similar to the pro's 24-second rule in colleges. He is opposed to such a move.

"Let athletes capitalize on whatever talents they have. It takes as much skill to perfect ball control as that needed to score 100 points a game," he says.

When asked if he could predict winner of the NBA title next season, the captain merely replied, "any doubt in your mind?"

Of course he was referring to the Los Angeles Lakers who acquired Wilt Chamberlain in an off-season trade to center a front line featuring Jerry West and Elgin Baylor.

LUCKY POODLE—Britt Ekland stars as Olimpia in the Warner Brothers film The Bobo.

Vets Make Best Directors

Saigon (MACV)—Officials at Red Cross national headquarters are convinced that ex-servicemen are among the best field directors and are actively recruiting them.

The buildup in Vietnam increased demands upon the Red Cross and created the need for more field directors. Following an initial orientation-type course, new personnel are placed around the states with Red Cross units for on-the-job practical training. At the conclusion of these preliminary periods they can expect to serve a one-year tour in Southeast Asia where 155 field directors are now handling an average of 22,000 emergencies monthly.

Primarily, these Red Cross representatives handle family and personal crises-caused problems for the servicemen and provide emergency communications with home.

Applicants may get more information from their local field director, by contacting Red Cross National Headquarters in Washington D.C., or by getting in touch with one of the following regional offices:

East: 615 North St. Asaph St., Alexandria, Va.; Midwest: 4050 Lindell Blvd., St. Louis, Mo.; Southeast: 1955 Monroe Dr., N.E., Atlanta, Ga; West: 1550 Sutter St., San Francisco, Calif.

Lieutenant Dinan returning the chicken head!

TWENTY-SEVEN

THE SHOTGUN PILOTS

IN the hierarchy of military clubs, ours were definitely in the lower-archy. We were small and therefore low budget. Entertainment groups, mostly from places like the Philippines, were in abundance throughout South Vietnam; had we had the resources, we could have booked a different act every night. As it was, we managed to afford about one booking per month, and these entertainment nights were always greeted with great enthusiasm. We got an occasional visit from a USO group, but in general, we were too little to attract the big stars. Happily, our hunger for entertainment was frequently met by a most unlikely and talented in-house duo.

Billeted in Eakin Compound were Captain Al Spain and Captain Paul Tanquay, 221st Aviation Company pilots whose jobs were aerial recon. They were referred to as spotter pilots, as their main mission was to spot problems and report. They flew small prop planes known as Bird Dogs, and although it was not authorized, they had affixed rocket launchers to the underside of their wings. This bit of macho was to benefit all of Can Tho during the Tet Offensive. Undoubtedly, someone was exercising his sense of humor when selecting their call sign, Shotgun. As a result, these two were known as and entertained under the moniker, The Shotgun Pilots.

Captain Spain was the son of a Southern Baptist preacher, and he had gained an ear for the art of fiery preaching. In civilian life, he had flown large cargo planes and expressed an enormous preference for cargo over passengers.

Captain Tanguay was also the product of a strict Southern upbringing and was reputed to have studied music under the tutelage of Ray Charles. As they roamed the skies over the Mekong Delta, secure in their Bird Dogs, they made excellent use of their ability to communicate with each other by re-creating the lyrics of various well-known musical triumphs so as to reflect the reality of what was happening on the ground beneath them.

Their performances would begin with a sermon delivered by Captain Spain. As all good sermons must be supported by the written word, the Captain would ask someone in the audience to provide him with a magazine. Quickly flipping through the pages, he would settle upon some article or advertisement and then, with no apparent hesitation, deliver a hellfire and brimstone sermon from the Book of *Time* or *Newsweek*, all the while making direct quotes from the written word. His sermons were certainly more amusing than admonishing, and as a rule, you had to strain to hear him over the laughter.

Then Captain Spain strummed his guitar, Captain Tanguay keyed the piano, and they sang their on-the-mark lyrics. The musical segment of their act always began and ended with their theme song "Strafe the Town and Kill the People." Somehow this little ditty spread to all U.S. military installations in Vietnam, and there exists today as many versions as there are for the world-favorite "Danny Boy." The lyrics they presented at that time were:

> Strafe the town and kill the people,
> Kill them in the village square.
> Kill them early Sunday morning
> As they're kneeling there in prayer.
>
> Throw some candy to the children
> And as they gather 'round,
> With your twenty millimeters
> Mow the little bastards down.

Much has been written about how lyrics such as these illuminated a certain blackness of heart said to have been all too common among the service personnel assigned to win the hearts and minds of the Vietnamese people. My on-the-ground experience assures me that all such psychological nonsense should be relegated to the trash bin of history. I lived with the performers, and I lived with the audience. This was all tongue-in-cheek parody designed exclusively for humor and stress relief. Thusly it was written and thusly it was received.

This talented duo concocted scores of parody lyrics, and had I in any way suspected that one day I might feel compelled to write about my experiences in the Mekong, I would have found a way of recording them. I didn't, and with the exception of the theme song, which was sufficiently repeated in my presence so as to burn into my memory, I recall only one other song in full. Borrowing liberally from the Beatles, the Tet Offensive inspired these memorable lines:

Yesterday,
All the VC seemed so far away,
But now I know they're here to stay,
Oh, I believe in yesterday.

Suddenly,
I'm not half the man I used to be.
The fear is locked inside of me,
And almost is the death of me,
Yes, I believe in yesterday.

Why we had to go
I don't know, they didn't say.
Now I know we're here to stay,
Oh, how I long for yesterday.

Yesterday,
War was such an easy game to play.
The VC seemed so far away,
Oh, I believe in yesterday.

Memorable lines from another inspired lyric modification were:

> Westmoreland's Cathedral,
> You're letting us down.
> You didn't do nothing,
> When Old Charlie came to town.

It went on, ending with these provocative lines:

> You can't say you didn't know,
> Because the Shotguns told you so.

Years later, when a big hullabaloo erupted over what the Saigon Command knew or didn't know about the VC build-up prior to Tet, I recalled that our spotters had spotted and had diligently reported what they had seen. I can't speak to what happened to those reports, but I can confirm that the Shotguns told them so. They truly were our "Eyes over the Delta."

TWENTY-EIGHT

CO SHUN

OCCASIONALLY I would visit the open-air market held on a large plaza adjacent to the Bassac River in the center of the city. Although the bulk of the goods arrived via the river, you could observe products being brought in by every imaginable transport, from little carts pulled by water buffalos or pushed by hand to modern trucks. What products, what produce, what unbelievable quality and abundance! Everything was so fresh, the brilliant colors virtually leaping off the charts. The area was so incredibly suited to agriculture that a farmer could expect three or four harvests each year for most varieties of lettuce and vegetables. When bringing tomatoes to market, the farmers would not waste their time picking and packing particular specimens. Rather, the entire plant was simply pulled out of the ground and brought intact—talk about fresh. I was particularly surprised by the presentation of the lemons put up for sale. These folks did not go through the trouble of picking lemons. No, a section of the tree branch would be cut, and the lemons would be presented on-the-limb!

It truly was remarkable that after all the years of conflict and killing, this food basket continued to overflow prodigiously. Happily, our tables were daily adorned with this extraordinary abundance.

Yet the most extraordinary abundance in any society is its people. This reality was impressed upon me many times during my stay in Vietnam. Unquestionably, one of the people who proved this maxim was Co Shun.

Co Shun was indeed a very special lady. She was, by force of nature more than by dictate, the head lady/office manager for the Can Tho Mess Association. Her English language skills were second only to those of Mr. Hein and easily exceeded those of many of our U.S. enlisted personnel. Always showing up at the office dressed in a very conservative yet spectacular traditional *ao dai*, she was, in the absence of Mr. Hein, the ultimate translator and problem solver.

I was uncomfortably surprised one day when she informed me that she would no longer be able to work for me unless I met with her father.

I responded in typical American fashion, "Not a problem, bring him in and I'll meet with him."

Evidently I missed the nuance altogether, and she, now somewhat embarrassed, was forced to explain that in order to meet with her father, I would have to come to his house and be formally introduced. I sensed her distress and readily agreed to make myself available at his convenience. I can't imagine what other response she might have expected, but my immediate acquiescence was received with genuine satisfaction.

On the day and time established for this important confrontation, Co Shun and I engaged a cyclo at the front gate of the compound, and under her direction, the driver delivered us to the walkway leading up to her family home. I was escorted into an ornate receiving room at the far end of which was seated a man who projected an aura of power and position. Co Shun had warned me that her father was very old-fashioned, but I really didn't understand exactly what that meant.

The gentleman was of undetermined age. His hair was white, he was attired in what I assumed was traditional apparel from shoes to cap, and he had the facial hair usually associated with the Chinese mandarins of old. He was seated more rigidly than relaxed, and he did not deign to rise. I could feel that I was being very carefully assessed as I was brought over to him. Co Shun said a bunch of things in Vietnamese that I did not understand, but I did pick out the words "*Thieu Uy* Dinan," so I knew that this was all part of the formal introduction. At length I was instructed to sit, and tea was brought in by a lady who I figured to be household staff. Co Shun remained standing just off to the side and acted as our translator in an exchange in which her father was clearly the questioner and I was the questioned. His questions were decidedly personal. Was I married? No. Then, did I have a lady friend? Yes. Was this a serious relationship given approval by her father? What university had I attended? Did I go to church? Which church? And so it went. All the while, his eyes kept boring into me, and I found myself sitting rather rigidly as well.

When tea and talk were done, he rose from his chair and began walking in measured steps toward the entryway. Co Shun advised me to follow, and I did. Never looking back, he continued through the portal and up the walkway to a waiting car with a uniformed driver. Personally opening the passenger door with his left hand, he simultaneously extended his right and, taking my hand in his, guided me into the car. When I was seated, he made a little speech that ended in the only words I could comprehend, *Thieu Uy* Dinan. Closing the door carefully but securely, he stood back and watched keenly as I was driven away.

The interview was unquestionably a unique experience, and I was happy to learn the next morning when I got to the office that I had passed muster. Co Shun was very happy as well, for had I not, she would have had to find other employment.

In talking with this unusual lady, I gained some interesting insights into the Vietnamese way of thinking on various subjects. A few of these insights revealed diametrically opposed cultural differences that have stuck in my memory.

On several of my trips to Saigon, I had noticed many young men, who appeared to be in the range of seventeen to twenty-one years of age, walking along in pairs and holding hands. No one around them, as far as I could determine, had ever given any of these duos a second look, but it struck me as an unusual practice. So I asked Co Shun, "What's the story on all these young fellows in Saigon walking around holding each other's hands?"

"Oh," she said, "that's nothing. When a young man goes off to the university, he is not allowed to have a girlfriend because it causes too many problems, so the students take boyfriends during their stay at university. Right now my brother is at the university, and he has a boyfriend. It's really a better way." All this was said without any hesitation, fudging, or embarrassment.

Maybe it was really nothing, but it sure seemed strange to me, so I asked, "What happens if, upon finishing at the university, a young man decides to keep having boyfriends instead of girlfriends?"

Co Shun smiled broadly and said, "Then we laugh at him."

Evidently, as far back as 1967, homophobia was not a serious Vietnamese concern.

Somehow a discussion about paying taxes came up, and Co Shun expounded with relish on various techniques that could be employed to avoid paying, if not all, at least most of them. I responded with the lecture about how a government couldn't operate without money, and therefore it was the responsibility of good citizens to pay their taxes so their government would be able to provide them with the services they needed. Assessing this high-toned moral position, Co Shun looked at me in such a way that I could tell she thought I'd lost my mind.

She went on to explain, "If the government were to use the money for the benefit of the people, what you say would be correct. But that's not the way it works. First of all, the tax collector puts as much of the money as he can into his own pocket. Then, when the politicians in Saigon get their share, they put most of it into their pockets. What do the people get? We get nothing. So why should we pay if we don't have to?"

"Well," I said, "if that's true, then people like you should get into politics and see to it that the people get their fair share of services."

"What do you mean by people like me? If I went into politics, I would do the same thing. That's the way it's always been done, and that's the way it always will be done here in Vietnam. We don't care about the government, and the government doesn't care about us. It doesn't really matter who is in charge up in Saigon. Whoever it is will do the same thing. They will take as much as they can for their own pockets."

Evidently there was little positive expectation on the part of the average

Vietnamese citizen where the central government was concerned. This may have been the root cause of our inability to rally the local populations around the Saigon leadership.

One day a check we had cashed for one of our soldiers came back to us marked unpaid because the account had been closed. This put my Vietnamese staff into a state of shock and disbelief. The person who had done this thing had committed an act of lying, and I was to learn that the universal understanding amongst the Vietnamese people was that Americans don't lie. They really could not believe it.

They told me, "This doesn't make any sense. This is lying, and you Americans are very strange, you don't tell lies. Whoever cashed this check had to be an American, and he lied. It doesn't make any sense."

All this genuine disbelief and outrage caused me to pose the obvious question: "What about you? Do you lie?"

Co Shun was insulted. With obvious indignation she replied, "Of course I lie. All Vietnamese people lie. You never expect someone else to tell you the truth. If you're too stupid to see through the lie, then it's your own fault, and you deserve whatever problems it causes you."

This certainly set me back and opened my eyes. I couldn't help asking, "Do you really believe that Americans don't lie?"

She responded with the simple statement, "Americans don't lie."

Now it was my turn to be shocked and I asked, "Then when you tell me something, should I assume that you are lying to me?"

"Oh no," says she, "that wouldn't be any fun. That would be like lying to a child. We would never lie to you."

Evidently, a number of her countrymen had other ideas of where to draw the line on speaking the truth, and a goodly number of our fellows had little understanding of the Vietnamese take on this subject.

TWENTY-NINE

SOMETHING IN THE WIND

I suppose that I should have been alerted to the reality that something somewhat evil was in the wind. One day, as I happened to be within the confines of the Eakin Compound Headquarters office, I was astonished to find myself confronted by a tubby little second lieutenant with the surname of Dewey, who had somehow summoned the gall to accost me. He had the cheek to demand that I sign receiving documents for a certain consignment of stuff ordered and received for my intended benefit. The stuff received, the receipt for which I was encouraged, yea veritably ordered, to affix my signature as the officer to become responsible for their allocation and use, were mess trays: those appalling things that resemble the aluminum containers used for frozen TV dinners. These horrid things were not destined to find a welcome place in my dining rooms.

Lieutenant Dewey was the compound supply officer, and his usual responsibility was to order such supplies as toilet paper and other necessities. So I asked him, "Lieutenant Dewey, why ever in the world would you take it upon yourself to place an order on *my* behalf without first consulting *me*?"

He explained that he thought that he was doing me the greatest of favors seeing as we were now a Class 1 mess and we were lacking this metallic marvel usual to military mess halls. "After all," he queried, "what will you do when you run out of plates to serve the food?"

I didn't bother to inject the fact that I had so many plates that this possibility was extremely remote. Instead I answered, "If I should need more plates I will get more plates."

"But," said he, in a gotcha tone of voice, "you don't have a budget for plates, and they're not in the military supply line." Happily he did not add "So there," but the ring of that attitude was heavy in the air.

I therefore informed him, "Should it come to pass that I am forced to make the choice between using those vulgar trays and reaching into my own pocket to purchase proper plates, I will definitely choose to make the purchase from my own funds. As for signing for those wretched things, I did not order them, and I don't want them. You ordered them, and you can sign for them and keep them in your own closet."

With those words I turned away and stomped out of the office, but it did cross my mind that there had to be more to this story than he was letting on. It was not to be very long before I was enlightened.

Colonel Hazen, the current deputy commander of operations, was one of those senior officers who believed that the real skinny, the gospel truth, was to be learned by consulting with the most senior members of the enlisted ranks. A couple of E-8s who, by the way, were scheduled to depart for home in short order, had taken umbrage at the sight of privates and other low-ranking personnel being served their meals on plates rather than being required to use a chow line and tray. Somehow they reckoned that this simple civilized activity was harmful to the long-term development of the proper military attitude, which should be instilled in our enlisted ranks as well as in the ranks of junior officers. It is difficult to assess their reasoning, as their reasoning was so very trivial. Basically, their position was that they had never been given such luxurious treatment and look at how wonderfully they'd turned out. Armed with such questionable logic, these sourpusses were able to convince the good Colonel that what we were doing was undermining the well-being of our troops and upsetting military order in general. As we now had a separate general officers mess, the conversion from served meals to mess lines would not infringe upon the lifestyles of the mighty and the proper way of doing things would be restored.

I was to learn all about this firsthand following another summons from Major Magnus inviting me to accompany him to Colonel Hazen's office for an important briefing.

When the Colonel let the cat out of the bag by telling me, "After much consultation with some very senior NCOs, I have come to the conclusion that we should abandon our current practice of operating dining rooms and start operating proper mess halls," this cat became all but paralyzed.

Having said that mouthful, the Colonel steeled back, evidencing some pleasure in the shock and horror clearly written all over my face, and awaited my response. Although nothing specific had been said, the recent mess tray fiasco bolted into the front of my mind; I was fairly certain that I knew where this was heading. I cast a hopeful look at the Major who in turn looked hopeless, but whose eyes had definitely widened, revealing a certain apprehension, knowing that I would not simply accept such a directive.

I tried to take the ignorant innocent approach, although having been blindsided, I was hardly prepared. I said, "Golly, I'm not sure I understand what you mean. I know that our operations could be considered somewhat upscale, and I think that the extra effort we put in is an important ingredient in the overall morale of the troops. I suspect that these very senior NCOs are the same ones who convinced you that we were poisoning the troops with our horrible food. I know that in response you made several unannounced visits for meals and concluded that the food we serve is better than what you get in the generals mess. I can't figure out how that's somehow improper."

With this he started to come forward in his chair; unkind words were clearly in the process of formation. Midway, he abruptly stopped, took a very deep breath, and then settled back once again. After a short pause, he said a surprising thing, "Forget about all that. It's time we move forward."

Now he gave me a curious look backed up by the slightest of smiles and made it obvious that I should be the next speaker.

"Yes, Sir?" I said, squarely putting the ball back in his court.

I could tell that this was not the response he had hoped for. You can generally tell when someone is trying to get you to commit to something independently so that they can later claim that it was your idea in the first place. I was trying not to go there, but the Colonel made another attempt to lead me.

In the most military of manners, he laid it out this way: "Lieutenant Dinan, I have made the decision that we should operate our mess halls according to customary military practice. It's up to you to tell me what you intend to do to implement that decision. Do you understand that, Lieutenant Dinan?"

"Certainly, Sir, but I think you will have to help me out on this one. You know that these are the only operations with which I have had any experience, and as far as I can tell, they are functioning very well." Again, this was definitely not the hoped for response.

Colonel Hazen knew that I was playing dumb, but at the same time, he couldn't prove it, so he pushed a little harder, "Lieutenant Dinan, are you telling me that you are not aware of anything unusual about your mess halls?"

I was beginning to enjoy this little charade, so I countered, "Well, I know that when the medical inspectors were called in, they reported that these were the cleanest, most sanitary operations they had ever encountered, so I guess that that's unusual but I can't recommend changing that."

He was not to be taken off course, and again he pushed a little harder, "Lieutenant Dinan, isn't there something else you find unusual?"

"Oh yes," I replied, and seeing that trace of satisfaction growing on his countenance, I threw another curve ball: "Do you realize that, considering the number of people officially stationed at Eakin Compound, we should probably max out at providing about 900 meals a day, and yet we've been serving 1,200 meals per day on a regular basis?" I knew that this tidbit of information was strictly off the wall and as far removed from anything he was hoping to hear as it could be, but since I was playing dumb, I thought it fitting.

He did not. "Lieutenant Dinan, that is another matter altogether, and we are not moving forward with ideas on how to implement my decision."

"Sir, may I make a recommendation?"

"That, Lieutenant Dinan, is what I've been trying to get you to do. What is your recommendation?"

"Well, it seems to me that you probably have some very specific ideas about how your decision should best be implemented, so maybe you could just tell me those ideas." This was definitely not the recommendation he was seeking.

Coming forward in his chair once again, he clasped his hands in front of him on his desk and gave me a look that I judged to be menacing. "Very well. It's hard for me to believe that you're actually this dumb. I was giving you the opportunity to show some real initiative and imagination, and you have failed to provide either, so I'll have to lead you by the hand. What you are going to do is this. You will reconfigure your operations so that a typical army mess line is provided for the benefit of my troops. Is that clear Lieutenant Dinan?"

"Perfectly, Sir."

"If you have any questions or something you wish to say on this matter, now is your chance."

I simply could not resist saying, "I want to go on record as saying that I think this is the worst decision I have ever heard, Sir."

This was not kindly received, and I was told, "You're on record. Now, get out and get to work." Naturally I stood and offered one of my smartest salutes; it was returned accompanied by one surprising word, "Out!"

Heading back to Eakin Compound, I endured another well-earned lecture from Major Magnus on what should and should not be said by a lieutenant to a full chicken colonel. As my Irish was up, up, up, I paid little attention and instead concentrated all my mental facilities on possible schemes whereby I might thwart the Colonel's intentions. When we arrived back at the compound, it was lunch time, and since I hadn't scheduled a tennis match, I went directly to the officers dining room for a solitary lunch.

I took the table in the far right-hand corner so that I could observe everything that was going on in the dining room. Although I ordered and ate, those two more or less automatic activities remained very secondary in my mind. Seated and surrounded by the relative elegance and comfort of this room, I became all the more overwhelmed by what a terrific operation I had inherited. I had not created this fabulous operation; that had been done by the likes of CWO Arnold and Sergeant Chavez. Yet I had worked very diligently, and with the able assistance of Sergeant Erickson, Mr. Hein, and others, I had succeeded in keeping it alive despite the various operational changes forced upon us. As I surveyed the happenings all around me, I reflected that this operation did indeed resemble a well-run restaurant more than it resembled the usual military mess hall. Eventually, I got up and walked around surveying every nook and cranny. I tried to imagine how I might introduce a mess line into this dining room, and my imagination failed me. I went into the kitchen and stood off to the side, observing the smooth if hectic flow of the cooking operation. More than anything else it reminded me of my experience in the kitchen at The "21" Club.

In a well-ordered kitchen, when the service of a meal goes into full swing, there is that peculiar drumbeat that resembles a locomotive charging down the tracks. It is unforgiving and unstoppable. It chugs and chugs, bah-bum, bah-bum, bah-bum, not stopping, never stopping, go, go, go. The peculiar intensity of concentration and physical activity endured by line cooks during the final production process is spectacular to observe, comparable only to an Olympic competition. Our athlete cooks were, without exception, gold medal contenders.

I sauntered into the enlisted dining room, which was larger by double-plus the size of the officers dining room and, although not as refined in some aspects, graciously appointed relative to customary military standards. Once again, I tried to conjure the possibility of replacing the dining experience I was looking upon with that of a chow line, and once again, my imagination failed me. Presented with such a dilemma, I asked myself the expected question, *Whatever in this world, should I, can I, will I do?*

In a word, I was conflicted. On the one hand, Major Magnus had graciously provided me with the proper military solution: "Keep in mind you're only a lieutenant. You have been given a directive from the Deputy Commander of IV Corps, who happens to be a full colonel. You should keep your mouth shut and do exactly as you have been instructed. It is not for you to concern yourself with the eventual outcomes. You have orders. You follow them." Somewhere within my thinking, I knew that he was, in every way, exactly correct. On the other hand, somewhere else within my thinking was a voice screaming: *Bullshit! These people are out of their minds.* This was my conflict. This required resolution.

I arrived at the mess office promptly at 1400 hours and checked for messages and other trivial things. Soon thereafter I informed my staff that if they should need me for anything, they could find me in my quarters. I knew they were anxious to find out what had transpired during my meeting with Colonel Hazen, but I simply was not ready to divulge anything about what I'd been instructed to do.

As always, just entering our well-appointed and gloriously air-conditioned living room lifted my spirits. I turned on our sound system, poured myself a cup of coffee, placed my bottom on our comfortable sofa, my feet on our coffee table, lit a cigarette, and engaged my mind. I briefly considered the idea of simply following orders. All too quickly, the horror and stupidity of those instructions again summoned up my Irish and all but obliterated my thinking capacity. Instead of seeing reality, I was seeing red, like an enraged bull ready to charge to his certain death. This was getting me nowhere. I had to calm down. I had to think. My thoughts turned from compliance to insurrection. There was no way that I could entice myself to lie down without a fight. But how does a lightweight go into the ring with a heavyweight and come out other than destroyed?

I understood that I could not win this fight on my own. As the pathetic reality of not having one single person I could turn to for help descended upon me, an alternate bizarre possibility entered into my mind. Although there was no *one* I could turn to, I could turn to *everyone.* I could duplicate the insurrection that had settled the grocery conflict. But how? How could I get the lieutenant colonels, the majors and the captains so stirred up that they would confront Colonel Hazen in such force as to have this decision reversed? In Army Intelligence they had taught us about a PSYOPS (psychological operations) technique wherein you feed people just the right amount of provocative information in order to excite their hunger to a point of desperation and then let them create their own feast. In civilian terms, it's called starting a rumor or coloring an upcoming event in such a manner as to evoke the maximum desired reaction, be that positive or negative. I spent the next couple of hours conjuring up the best method of provoking as much discontent as possible.

At length I settled on this ice-breaking opening statement: "I hope you're prepared to start using a mess line." To my way of thinking, although it wasn't really a question, it was a statement so ripe with provocation that a questioning response would be irresistible. I was to be proved correct beyond my wildest imagination.

I got myself to the officers club sometime prior to the opening bell at 1700 hours and awaited my first victim like a hungry feline in ambush. I restrained myself sufficiently to allow the first fresh meat to mellow through half a drink. Attempting to sound as ambivalent and nonchalant as possible, I approached this unsuspecting major and in the most matter-of-fact way said, "I hope you're ready to start using a mess line."

The victim first looked me in the eye rather strangely and then looked to my shoulder in order to determine what level of idiot had made such a stupid statement. Bringing his unbelieving eyes back in line with mine he asked, "Lieutenant, whatever in the world are you talking about?" This was perfect.

"Well, Sir," I replied, "the fact is that a couple of disgruntled senior NCOs, who incidentally are about ready to ship out, have convinced Colonel Hazen that we should get rid of our waitresses and institute a chow line like those in usual mess halls."

There is disbelief and there is disbelief. This was true disbelief! If ever you have seen that particular expression on another man's face, the one that conveys more succinctly than any words the fullness of his disbelief, then you have seen the face into which I was looking.

"Give me that again, and tell me exactly what the hell it is that you are talking about." Who could ask for anything more? Given this directive by a senior officer, I was compelled to deliver all the gory details of that morning's meeting. In the end I was left with the statement, "We'll see about that."

I went from pillar to post. I worked the bar, the booths, the tables, the pool room, the lounge, and anyplace else I could find an officer's ear available to my attack. Soon there was electricity in the air, a murmur seeking to become a roar. Anger, that very special emotion, was spreading and billowing amongst this select group of unsuspecting victims who, feeding on one another's determination, declared their determination not to be victimized. I tumbled exhausted into my bed, not knowing what the morning would bring. Yet I knew that I had succeeded in starting an insurrection.

I arrived at my office with more than usual punctuality, pungent with high expectations of learning something of the results of the insurrection I had inspired.

I always had more paperwork on my desk than I could have imagined. This ongoing task served to reinforce my understanding of the military dictum: "The job's not over until the paperwork is done." It should be noted that everything, everything, everything done in the military requires full and extensive on-paper documentation. The responsibility of executing this mandate is placed upon junior officers. It was assumed that the actual activities upon which this documentation was predicated would be generated and overseen by various senior NCOs who were armed with considerably more hands-on military know-how than could be reasonably expected of an infant officer. The fact that I actually took charge of the operations under my auspices in no way diminished this paperwork responsibility or liability. So, disposed or indisposed beneath the usual paper monster, I scarcely noticed the ticking of the clock until I was removed from the comfort of my labors by the uncomfortable words delivered through my telephone, ending with the admonition, "You will hold for Colonel Hazen."

With no small degree of trepidation, I held on as ordered. Without sufficient time to properly gird my loins for the coming onslaught, I was assaulted by the bombastic and blistering voice of Shazam screaming into the phone, "Goddamnit, Lieutenant Dinan! You did not listen to, or you did not understand, my orders! I never intended that you should eliminate the waitress service from the officers mess. My orders were in reference to the enlisted mess *only.* Do you now understand what I want you to do?"

"Yes, Sir."

"Then do it, goddamnit, and keep me the fuck out of it. Do I make myself clear?"

"Yes, Sir."

Bang went the phone. Bing went I. Uncharacteristically, I leapt from my chair shouting, "We won, we won, we continue on course!"

This stupefying announcement and its accompanying exuberance were met with total bewilderment by both my Vietnamese and U.S. staff alike. In unison, they turned their eyes upon me and delivered a soundless verdict that could only be interpreted as "This *Thieu Uy* has lost it once again." I was exultant; I couldn't have cared less. It was close to lunch time, and I was far too energized to do anything energetic.

Casting my glance upon all those uplifted and bewildered eyes, I announced summarily, "I will explain after lunch."

I retreated immediately to my quarters and concocted a splendid lunchtini—that's a martini, which one should never, except in instances of extreme victory, even think about drinking prior to the accepted cocktail hour. Suitably exhilarated and half-buzzed, I found my way to the officers dining room seeking commentary on what activity had inspired the Deputy Commander of IV Corps to personally contribute so much joy into my life. I was not to be disappointed.

Somehow, the word had been immediately communicated to all interested parties that the status quo would be maintained as regards the OM. The inspiration for this turnabout on the part of Colonel Hazen was the assault upon every G section chief by his supporting officer corps regarding the bizarre decision to denigrate them by forcing them to subject themselves to the humiliating specter of a chow line. As the story went, lines formed at the offices of G this, G that, and G everything, comprised of lieutenant colonels, majors, and captains demanding to know why their dining experience was going to be downgraded because of the ruminations of a couple of demented NCOs who were about to return to the States. In turn, these senior officers, who had been deprived of any knowledge of what was going on by virtue of being sequestered within the confines of the rarified sanctum of the generals mess, descended en masse upon the person of responsibility, the Deputy Commander of IV Corps, Colonel Hazen. Confronted by this august assemblage demanding reasons why their officers were to be so abused, the Colonel ignominiously capitulated. He then placed the aura of responsibility for such a ridiculous concept squarely upon the shoulders of that misunderstanding reprobate, Lieutenant Dinan. Although no one believed this augmented account of what transpired, no one cared or dared to sunlight the truth. Civility was preserved, happiness prevailed, and life would go on. Lieutenant Dinan, that's me, would have to look after whatever fallout followed on his own.

I, young and stupid, couldn't have cared less about all the political ramifications attendant to this outcome. I had won the first skirmish, and now I concentrated my energies on the prospect of winning this conflict absolutely, completely, and entirely.

I assembled all my enlisted staff and reported to them all the gory details of everything that had transpired. They were aghast to learn of the betrayal perpetrated by their enlisted brethren. They were blown away by the fact that the officers would escape the consequences of this stupidity while they were to be the victims. My guys were not simply angry, disappointed, or mad; they were enraged, invigorated, and fighting mad. They demanded to know what they could or should do. I explained that I couldn't be involved but, if they were so inclined, they might consider the idea of revealing what they knew to their fellow soldiers and encouraging them to voice their discontent to their seniors in the NCO ranks.

Like a well-organized counterintelligence network, my men spread out and attacked or assaulted every ear that their lips could find. They worked the enlisted dining room, the EM club, the senior NCO club, the shower stalls, the gathering spots, the craps table, the whatever, anyplace, anywhere. They orchestrated such a crescendo of malcontent that the bile was overflowing like the waters of the Niagara before the last of our enlisted men placed his head upon his pillow.

At sunup came the showdown.

The two great leaders and exemplars for the enlisted ranks, our compound's First Sergeant and the Fourth Corps' Sergeant Major, were everywhere attacked, surrounded, questioned, and hassled in their attempts to traverse Eakin Compound. Such was the multitude of the supplicants that surrounded them, an observer would think they had assumed the status of rock stars. Closer inspection would reveal that they were surrounded by angry mobs simply demanding satisfaction regarding their dignity as diners.

The music played on, and at 1030 hours, when I answered my phone singing the usual stanza, "Can Tho Mess Association, Lieutenant Dinan speaking, Sir," I was rewarded by what was to be music to my ears.

The very voice of Colonel Hazen, lacking any melodic content, pierced my hearing with the, to me, beautiful musical lyrics: "Lieutenant Dinan, I don't know what the fuck is going on over there at Eakin Compound, but you will keep your fucking waitresses and all the bullshit that goes along with whatever the fuck it is you are doing there, and I don't want to hear another fucking word about this crap ever again as long as I am in charge. Is that fucking clear?"

Tut, tut, such language, and from a senior officer. I, for one, was not offended and I responded in the finest military manner, "Yes, Sir. Thank you, Sir."

At that moment the phone went dead, and we celebrated life as I jumped up and screamed, "We did it! We won!"

The power of the people: how fabulous, how dynamic, how certain, how satisfying. This power should never be underestimated.

THIRTY

USAID

FOLLOWING this stunning victory over administrative silliness, malfeasance, malfunction, maliciousness, misguidance, or mischievous stupidity, I mistakenly assumed that I would be exempted from further such incursions from on high, well described in military jargon as SNAFU (Situation Normal All Fucked Up). How wrong can one singular lieutenant be?

I was still basking in the blaze of victory while plotting, planning, and perspiring in the ongoing effort to maintain and augment our hospitality support system within our little segment of this noble conflict, when I received a most unexpected phone call. Having answered my phone with military courtesy in my customary manner: "Can Tho Mess Association, Lieutenant Dinan speaking, Sir," I was alerted to the probability of something sinister when I heard the disquieting words, "May I introduce myself?"

This particular opening statement is recognized the world over as that used by someone trying to sell you something that you do not want.

My first thought was *No, you may not; just please go away,* but again, in keeping with military courtesy I replied, "Yes, certainly."

Once given this expected green light, my caller, demonstrating a voice radiant with confidence and position, proceeded to assault me with the most galling and appalling demands. First came the self-introduction: "I am Mr. DePew. I head USAID activities in the Delta."

The pause following this admission was held long enough that even I recognized that some acknowledgement of this salient fact required verbalization, so I responded, "Yes?" With malice and forethought, I left out the *Sir* and injected as much question into that single-worded response as my diction would allow.

The ensuing elongated pause permitted me to conclude that my intended message, amounting to *So what?* was, if not well, certainly received. I was pleased to note that when my caller continued, his voice had lost that edgy inflection particular to people imbued with boundless authority that causes most thinking persons of lower estate to submit to their presumed overlords.

Before continuing, two things should be noted: One, sometime prior to this call, I had, at Mr. Hein's instigation and with his able assistance, revised the payment structure and benefits program for the personnel employed at Eakin Compound. The results of our combined activities placed the Can Tho Mess Association in the position of being, at that time, the most generous and progressive of all U.S. military employers anywhere within South Vietnam. Two, at the time of my telephone call from Mr. DePew, I had no sense of just how important a personage he was.

And so this Mr. DePew went on to describe many of the measures he intended to inflict upon my operating procedures regarding the Vietnamese personnel employed by our mess association. He gave me the rundown, in great detail, as to how I would submit to his authority and organization in all matters concerning the hiring, disciplining, and possible firing of any Vietnamese personnel employed by the Can Tho Mess Association. To this end, I was to turn over to his office all records concerning the medical examinations and other pertinent personal documentation of all my Vietnamese staff for his evaluation. Further, in the future, I was to contact his office in the event of any hiring, firing, or disciplinary needs and to comply with his assessment of what might be the correct course of action in each instance. With regard to all this, I was given his personal assurance that this was not only for my benefit, but for the greater benefit of our Vietnamese allies and the entire civilized world.

He would look after all my needs. I would simply submit all my personnel-related decisions to his discretion, and he would, in his own person, become the answer to any and all my attendant problems. I was staggered by his confidence and enthusiasm. His evident recovery from our initial exchange was manifest. I scrupulously considered his supposed power and purpose, and with little regard for either, I articulated the thought that was foremost in my mind.

I said, "You're out of your mind." Having delivered this cogent evaluation, I gently hung up the phone.

I suffered no illusion that this would be the last I would hear on this front, but I suspected that I had, at the very least, bought myself some time. My head was swimming as I asked myself the oft-repeated question, *Why can't they leave me—and well enough—alone?*

In record time, my phone rang, and I was confronted by the clearly agitated voice of Major Magnus, demanding: "Have you any idea what you have done?"

Well, I knew perfectly well that I had done a bad thing. I was not, however, aware of just *how bad* a thing. Struggling for an answer that would best serve my interests, I came up with the lame response, "Could this have anything to do with a Mr. DePew?"

I could hear the Major sucking air before he, not very gently, assured me, "You're goddamned right it has to do with Mr. DePew. Do you have any idea who he is?"

"I believe he is a labor organizer or something like that."

This inspired the Major to retort, "I believe you had better get your ass over to my office, and get it over here now." As the phone immediately went dead, there was nothing else to say.

As I could find no way of disengaging the requested body part from the rest of my person, I took my full self immediately to Compound Headquarters.

Upon entering I was greeted with looks laden with compassion. Words would have been redundant. The full body language of every man present fairly screamed, *Oh boy, what has Lieutenant Dinan got himself into this time?*

I attempted a cheerful demeanor and gave a "Hi guys" salutation to the assembled.

The First Sergeant smiled knowingly and said, "You may go right in. Major Magnus is expecting you." So in I went.

The Major was seated at his desk looking thoughtful and glum. When he looked up at me, he disarmed me completely by presenting me with that curious look generally reserved for a loving parent who is forced to confront a child who has deeply disappointed him.

He said simply, "Take a seat."

I sat ramrod straight in the finest show of military respect and said, "Thank you, Sir."

He just looked at me for a little while, not unkindly but rather very thoroughly, as though he was seeking something very specific. When he was evidently satisfied with this optical investigation, he began: "Lieutenant Dinan, I know you work your ass off, and I am convinced that you're doing a great job with all the responsibilities that have been heaped on your shoulders. But sometimes I think you must have more than a few screws loose. I have had occasion to caution and reprimand you regarding things you've said and attitudes you've displayed to your superiors more times than I like to recall. Now, let me tell you about Mr. DePew, who found it necessary to call me because of your rudeness."

I understood that he was not really asking for my permission and that to give it would have been insulting. There was something in me that wanted to say *Please go ahead; I will let you.* Happily, greater wisdom prevailed. I kept my mouth shut and sat up a little straighter in a show of respect and expectancy.

Looking at me calmly and squarely, Major Magnus continued, "Mr. DePew happens to be *the* senior ranking civilian in the Fourth Corps. Mr. DePew runs the United States Agency for International Development operations here, and every civilian operative is under his command. Mr. DePew has the protocol rank of a lieutenant general. Can you tell me how many stars a lieutenant general has?"

"Three, Sir."

"Very good, and can you tell me how many stars our Commanding General has?"

"Two, Sir."

"Very good again. Now, putting all this together you will note that, from a protocol standpoint, Mr. DePew is not only the senior civilian in IV Corps, but also the senior American representative of the United States. Do you think it is a good idea to hang up on the senior representative of the United States?"

I supplied the only reasonable answer, "No, Sir."

"Very good; we may be getting somewhere. So tell me, do you think it is proper to tell the senior representative of the United States that—what were the exact words? Oh yes, here it is, 'You're out of your mind.'"

"That would not be proper, Sir."

The Major was now on a roll and he asked, "Do you recall receiving a phone call from a gentleman identifying himself as Mr. DePew?"

"I do, Sir."

"Do you recall telling that gentleman 'You're out of your mind' and then hanging up your phone?"

"I did both those things, Sir."

"Then we have a little problem to deal with don't we, Lieutenant Dinan?"

"Sir, I think I can explain."

This was evidently the wrong answer because this time, instead of saying, "Very good," Major Magnus said, "No, no, no. There is no explanation. Nothing you can possibly say can alter the fact that what you did was wrong, wrong, wrong. Had you done this to a military officer of similar rank, you would be court-martialed on several counts, including insubordination and disrespect."

I was starting to get a little fidgety because I wanted him to hear my side of the story. This did not go unnoticed, and the Major assured me, "I can tell that you have something you dearly wish to say. When you leave my office you may find a wall and tell that wall anything you wish. I don't want to hear any of it. Is that clear?"

"Yes, Sir." (What else could I say?)

"Very good. Now this is what you are going to do. I have scheduled a meeting with Mr. DePew at 1500 hours here in my office. You will report to this building at precisely 1515 hours. You will take a seat and wait to be called. When you are called, you will be formally introduced to Mr. DePew, and you will immediately make a full and sincere apology for your outrageous behavior. You will not make an explanation. You will make an apology. Is this clear?"

"Yes, Sir."

"How clear, Lieutenant Dinan?"

"Very clear, Sir."

"Very good. Now get out of here and go find your wall. And don't fuck this up."

There was nothing left to do except leave, so I did. I didn't bother with the wall thing. I simply went back to my office and did mostly nothing for a while and then went to lunch. I was suffering somewhere between humiliation and rage, and I felt badly for putting Major Magnus in an awkward spot. Also, I was feeling nervous about my upcoming meeting, and I was determined to handle whatever came my way with as much grace and dignity as possible.

Time in these circumstances is never found to be a slow-paced friend, but rather a fast-paced foe rushing you to certain disaster. At 1515 hours, as ordered, I was miserably seated in the outer office like a schoolboy outside the dean's door awaiting the inevitable chastisement. I was allowed to cook and stew. My friend Time now slowed to a snail's pace. At 1535 hours, the inner door opened, and I was summoned within by Major Magnus. As I crossed the threshold, I observed a man, wearing a medium gray suit of excellent cut, rising slowly from the right-hand guest chair. As he unfolded, it became obvious that he was tall, six foot two or three, and imperially slim. When he turned toward me, I immediately took note of his highly starched collar and subdued regimental tie. A slightly stooping posture completed the very picture of a highly successful career diplomat. Major Magnus made the introductions, and Mr. DePew took my hand firmly but not aggressively. I wasted no time in gushing out my apologies, which were accepted with great nobility, accompanied by the assurance that responsibility should be shared in light of my not having been given the benefit of a full and proper introduction.

At this point, everything seemed to be hunky-dory, and Mr. DePew let it be known that he was anxious to restart our conversation regarding my Vietnamese personnel. I made the suggestion that it might better serve our needs, while freeing up the Major's office, if we moved our conversation to the mess association office. This was met with all-around approval, and I graciously escorted this dignitary back to my office with the best of intentions for a productive and amenable outcome.

In short order, we entered the mess office, and I introduced my guest to one and all. Curiously, I was immediately struck with the impression that, unlike my American comrades, my Vietnamese cohorts knew exactly who this dignitary was, and something in their demeanor signaled a singular dislike and distrust toward him. It was nothing I could specifically put my finger on, just a general impression that I couldn't quite shake or comprehend. I had learned to trust their instincts and judgments, and as a consequence, the caution light that was at all times dimly burning somewhere within the recesses of my Bronx-infused mentality lit up and put me uncomfortably on edge.

As soon as we were seated at my desk and supplied with coffee, Mr. DePew began his pitch. He carefully explained how important it was to our overall mission that we treat our Vietnamese civilian employees with kindness and respect. He brought out charts and other documentation proving the disparity of payment practices prevalent throughout the many military installations in the Fourth Corps. He detailed for my benefit numerous cases of managerial malfeasance associated with the discharge of various civilian employees. He provided me with detailed minimum wage scales to be applied for every conceivable level of employment. He described the various funds that we would have the opportunity of contributing to for the ongoing benefit of these benighted locals. He extolled the benefits of a central hiring system that would relieve me of all the burdens associated with the dire problems attendant to securing the qualified, sanctified, and purified laborers necessary for our very survival. To merely acknowledge that this executive was forceful and well prepared would diminish reality. This Music Man had all Seventy-Six Trombones tuned to a perfect pitch.

While absorbing his slick, professional, and heavily documented presentation, I had the distinct advantage of being so situated that while I could observe every nuance of silent commentary generated by my Vietnamese allies, my guest, with his back to this audience, had his visual sphere limited to me and the unappealing corner of our office occupying the space beyond my desk. The usual office chatter was, in a word, not. Every available ear strained to catch every syllable. I surreptitiously engaged the speaking eyes of Co Shun whenever prudence allowed. I was rewarded with such an earful that the notion that one must talk to speak no longer struck me as a self-evident truth. The unspoken message could not have been more clearly or loudly articulated: *Don't be a child; think like an adult.* I was conveniently and covertly forewarned and forearmed.

When his full, perfect, and sufficient presentation came to its inevitable conclusion, Mr. DePew settled back comfortably and politely inquired, "Before I present you with the usual papers for your signature, do you have any thoughts or questions that should be considered?" At the same time, the barrage of wordless vindictive discharged by my indigenous staff was so overpowering that I could now physically feel it without the benefit of any eye contact. In my own way I tried to ignore this onslaught, but that yellow light brightened considerably.

I reviewed for the enlightenment of Mr. DePew the fact that each wage and benefit detailed on his graphs and lists was paltry in comparison with the standards we at the Can Tho Mess Association had already established. He immediately acquiesced and assured me that this discrepancy would in no way cause a problem. Further, he would personally see to it that our generosity would be protected by an addendum attached to our contract specifying that we would continue to pay our personnel at the favorable levels they now enjoyed, which were considerably higher than the level of wages and benefits established by his office. Somehow, I was less than overwhelmed by this declaration of helpful cooperation. I proceeded to outline my misgivings regarding subcontracting control of hiring and other staff related management activities to an outside power. This concern was preemptively dismissed with the argument that whereas *I* would ultimately be gone, *his* organization would remain in perpetuity, thus ensuring the continued well-being of all concerned long after my departure.

His arguments were strong, and with recognizable glee, he placed before me a folio of documents for my signature, at the same time letting me know, "If you will now just sign the documents I have prepared for you, we can conclude our business, and you and I can attend to our other pressing needs."

Thus confronted, I moved the documents, without giving them the slightest inspection, to an empty spot on the right-hand side of my desk. This accomplished, I informed my guest that I had made a covenant with myself never to put my signature to anything in the presence of the requesting party. In an attempt to appear both convivial and cooperative, I then told Mr. DePew, "After I have had the opportunity to carefully review the papers you have so very kindly prepared for my benefit, I will contact your office regarding my decision."

Upon hearing this businesslike proposal, an incredible transformation came over this person sitting in my face. Suave gave way to sweat, and asserting his long right index finger by jabbing it on top of the put-aside documents, this overtly distinguished diplomat informed me in no uncertain terms, "I am not leaving this office until you have signed these papers."

I suspect that I was meant to be intimidated by this demonstration of executive determination. I suspect that Mr. DePew had no inkling of the yellow light penetrating my thoughts. I suspect that this unfortunate civil servant had no knowledge of the silent zingers aimed by Co Shun. I suspect that this gentleman had no way of recognizing that my yellow light had suddenly turned to red.

How does it happen? In that very instant, my eyes glazed over, and all caution was cast to the winds. I stood and with very formal dignity announced, "Sir, you are entirely wrong. You will quit these premises immediately."

I cannot precisely describe the reaction this declaration caused. Utter and total disbelief is as close a description as possible. But it was far more complex: anger, dismay, insult, disbelief, fury, outrage, and rage. When Mr. DePew at last galvanized himself sufficiently to enable his exit, I suspect that our tempers were on equal footing.

His parting words hung over me for a protracted period of time, "You have not heard the last of this; you shall hear from me again."

For all my bravado, I was a complete wreck upon his departure. Co Shun was instantly at my side assuring me, "He no good, you done right." Who knoweth? I never heard another word from or about Mr. DePew, but I was later advised that prior to my departure, my replacements, four captains all told, all signed up as beneficiaries of this administrative boondoggle.

THIRTY-ONE

HIGH JINKS AND HIGH REGARD

LIFE moved on in Eakin Compound, and the annual monsoon, famed for its aerial onslaughts of both torrential rain and tormenting swarms of flying insects, moved in. The rainy season in the Mekong Delta must be experienced to be thoroughly appreciated. Twenty-four hours a day, for days on end, the skies provided an unrelenting downpour of therapeutic intensity. The volume and consistency of this unwelcome watering can be fairly well imagined if you call to mind the greatest showering experience of your life. You remember, the one where the water pressure remained so strong and so constant that it beat every pain from your well-wearied body.

The regrettable difference was that instead of being unclothed in a private space intended for the enjoyment of a splendid, self-inflicted torrent, you were of necessity fully clothed, in public, and compelled to move about as if the sun were shining. As a result of this circumstance, you became, with neither your intent nor permission, wet. Everything became wet. Damp soon replaced dry as the hoped for achievement on the moisture scale. The all-encompassing, and for us unprecedented, dampness inflicted a skin irritation we as a rule associate only with the very young: diaper rash! This natural enemy claimed victory over multitudes of military men. Lacking sunshine, moonshine—or one of its more sophisticated derivatives—helped to soothe much of the discomfort. Undoubtedly, an equivalent external application of talcum powder would have accomplished greater relief, but the macho mindset of the time generally rejected that option.

Thus wetted down externally and internally, the American military might stationed in the Delta sloshed through the rain. Whatever hope one had of experiencing joy during the brief interludes of actual sunshine was diminished by the clouds of obnoxious and bloodthirsty flying insects that then filled the skies. Through the medium of inflicting us with these airborne irritants, Mother Nature maintained her position as dominatrix: while the insects feasted on our flesh and blood, she knowingly sopped up moisture in some far-off clime with the specific intent of our future victimization. The overwhelmingly superior magnitude of Vietnam's insect population outclassed anything I have ever experienced or heard about. This phenomenon may well be testimony to some positive ecological dimension peculiar to the Mekong Delta. Remaining ignorant on that esoteric level, I can only report that living through those infestations was, for this soldier, in every dimension repugnant. Nonetheless, in the finest tradition of military heroes, we prospered and survived, damp and dripping, with dignity.

In contrast to our prolonged survival mode caused by nature's unalterable malevolence, it was that alterable atmosphere or pressure conjured and controlled by the minds of men that was to give us concern about our survival. Things began to change in many little ways. So subtle and indefinable in their appearance were these little indicators of impending doom that they failed to alert the rank and file in sufficient measure. It was just so many little things; nothing you could put your finger on.

The two somewhat measurable indicators were an increase in military population and a greater insistence on regular military protocol. Along with a noticeable increase in U.S. military personnel assigned to serve in this adviser stronghold, an undercurrent of that *snap to* mentality began to insinuate itself into every aspect of our daily lives. But again, this undercurrent of change was so subtle that although it became pervasive, no particular action or aspect can be singled out as an example worthy of mention. In civilian life the same experience is often exemplified by rank and file workers noting to one another,

"I don't know what's going on around here, but something's not right." The reality of the increase in military personnel, however, was plainly evident, and the impact of this increase on the occupants of C-5 was quick and decisive.

As mentioned earlier, by dint of harboring the only resident chief warrant officer, we in C-5 were blessed in the unique luxury of our air-conditioned quarters. We also had the distinction of being the lowest ranking grouping of officers billeted in Eakin Compound. In recognition of this distinction, we were the first to receive orders to increase our occupancy from three to four. Our quarters consisted of a two-room suite. One room served as our living room or lounge, and the other our sleeping quarters. (It would be far too misleading to describe this as a bedroom.) The arrangement of the three beds was one freestanding and two bunked or stacked. The simple solution to the necessity of accommodating a fourth roommate was clearly to simply stack another bed on top of the freestanding bunk. In civilian life, a bunk bed is understood to be two beds, one stacked on top of the other. In the military, every conceivable platform designated as a sleeping space, no matter how contrived, is a bunk. The snag in carrying out this self-evident and simple solution was that our beds were *beds*. That is to say, they were real full-size single beds as opposed to the standard army cot. A surplus of army cots was on hand. There were no surplus beds. What to do?

We held an emergency meeting of the board of directors of C-5. Well, anyway, we three roommates gathered for a protracted discussion on how we might best deal with this frightful dilemma. As we enlightened our best thinking with our best and favorite beverages, our clarity of thought reached unsuspected and lofty plateaus. Our opening dialogue centered upon how to determine which one of us would be subjected to the indignity of an army cot. There was to be no walking out on these negotiations, so it was imperative that a compromise acceptable to all parties be agreed upon. With carefree thought and self-serving deliberation, it was mutually agreed that although there were no unclaimed beds suitable for augmenting our freestanding bed, thereby making it a double bunk, this coveted piece of furniture would simply have to be found on post and then find its way into our quarters. Having thus clearly determined and defined what had to be done, the objective, or end goal, of our mission was set.

As described earlier, Eakin Compound was divided by a central parking lot with one side designated for enlisted personnel and the other for officers. There were no beds allocated for use on the enlisted side of the compound. Even so exalted a personage as the Sergeant Major of IV Corps suffered the indignity of sleeping upon a humble army cot. Possessing this vital intelligence, we determined that the reconnaissance phase of our mission would have to be limited to furniture found within the commissioned officer stronghold. This was indisputably a zero-sum game. For us to win, someone clearly had to lose. How to choose? Who would be the victim? The sobering question *How can we possibly get away with this?* failed to find its way up to our lofty plateau. At length a cogent plan was formulated and agreed upon.

The next day I cornered my club sergeant, Sergeant Cole, in his little office within the enlisted club. This was a particularly apt place to communicate for-your-ears-only instructions, as this windowless workspace was well within the facility and not subject to passing by cowans and eavesdroppers. Typically, a military club sergeant will transgress the strict rules that guide his occupational specialty from time to time. Happily, Sergeant Cole was typical, at least in this context. Providently, he had provided me with the opportunity of bailing him out on more than one occasion. Thankfully, I had risen to those occasions, and therefore he owed me big time. It must be understood that certain delicate transactions can never be effectively perpetrated by duly commissioned officers. The arena of discreet underhanded activity is recognized to be an exclusive domain of non-commissioned personnel only. It is imperative that in the event that an officer finds need of discreet underhanded activity, the entire onus, should the activity come to light, must be directed squarely and solely upon the enlisted co-conspirator. Bearing this dictum in mind, I did not tell Sergeant Cole what to do. Rather, I simply communicated a need.

With discretion and in absolute privacy, I stated my need: "Sergeant Cole, I need a fourth bed installed in my quarters. Tomorrow, between 0900 and

1600 hours, no one will be in C-5. When we return after 1600 hours, a fourth bed will be duly installed. Have you any problems with that?"

He uttered one word, "*Bed?*" This singular response delivered in question format signaled that he understood the difference between bed and bunk.

I confirmed, "Bed."

Sergeant Cole looked like he had been hit with a brick. Of course there were problems; he knew it and I knew it, but the way the system worked, he was obligated to assume macho, can-do posturing. He postured very well under the circumstances and with hardly a flinch came right back at me with this tantalizing question, "Will this even us up?"

I hesitated a moment as though I really had to think about it, and implying that he had gotten the better of me in this exchange, I replied, "Well, OK. We'll call it even."

Although the challenge I had laid down was demanding, and in a way not very nice, I was confident that he would be up to the task. The installation would be a simple matter inasmuch as we had two full-time carpenters on the payroll who regularly worked under Sergeant Cole's direction. The ticklish part of the operation would be securing the bed in the first place. This was the discreet activity concerning which my complete and uncompromised ignorance was paramount.

The following afternoon when my roomies and I returned to C-5, we were delighted to note that a fourth bed had spontaneously appeared, duly installed. As we were crossing the volleyball court on our way to the officers club with celebratory intent, we were shocked, shocked to observe a senior major, one of the select few who commanded single-occupancy quarters, burst forth from his coveted quarters. The man seemed to be positively deranged.

Red-faced with rage, his arms flapping all over the place like a wounded crane, the man kept screaming, "My bed, my bed, my bed," all the while running about in senseless circles. Such a display by a senior military officer—tut, tut, can you possibly imagine? We celebrated indeed.

As to "How can we possibly get away with this?"—I don't know the answer, but I can categorically state that we did.

Driven by a daily grind, replete with operating activities that remained ongoing twenty-three hours a day, seven days a week, my time fairly flew by. This time warp was accelerated by the monthly requirements of board meetings, minutes writing, production of financial statements, trips to Saigon for currency exchange, and the monthly round robin with the various G sections to explain every questionable detail before the monthly documents landed on the CG's desk for signature. However, the most wearying and time-destroying activity centered on dealing with the insistent bitching.

It is a universally accepted maxim that no matter how diligently you try, no matter the quality of the product you deliver, in the service sector, no one bats a thousand. "You can't fool all the people all the time," is a well-stated reality, well understood and respected by all. The analogous statement, "You can't please all the people all the time," fails to garner the same acceptance. People couldn't care less about all the people being pleased; their concern is concentrated on having their own particular idea of pleasure satisfied. In the real world, the displeased takes his purse and person to another place. In the make-believe world inhabited by an overseas army where there is no other place to go, the displeased take their disappointment to their lips. Ironically, the American military has a saying: "If the troops aren't bitching, then something is definitely wrong." By that accounting, everything we were accomplishing at Eakin Compound was definitely right or, at the very least, other than definitely wrong.

Bitch, bitch, bitch. The more senior the bitcher, the more sonorous the bitching. It was this constant drone of malcontents that drove Sergeant Butler to the point of taking his life. And it was this very same constant drone of malcontents that placed me on a bench one evening shedding tears upon the benevolent shoulder of my board chairman, Colonel Barnes.

How does it happen? How do all the frustrations, conflicts, difficulties, and bitching coalesce at one moment in time in such a way as to shatter one's confidence, determination, and bravado? Who knows? All I know is that it happened to me, and I can't identify that single straw that broke this camel's back.

It was just another usual day, nothing special to report. I had done all the usual things in the office and about the compound. I had had the usual fine dining experience at dinner time. I was in my quarters engaged in the usual contemplation of tomorrow's most pressing needs for my attention, when I found myself overcome by a sense of despair. I felt tired, picked upon, put upon, unappreciated, unrewarded, unwanted, and altogether dispirited. I needed to be alone, so I went outside and sat on one of the benches along our walkway and looked up into a spectacular star-filled sky. It was one of those rare and special evenings when the skies were cloudless and bugless; the air, balmy and breezy. My external comfort did not sop my internal misery. I was so consumed by thoughts of my trials and tribulations that there remained no space for the inclusion of my triumphs. I was so thoroughly engaged in the nonproductive, self-destructive process of self-pity that it must have cast an unwelcome aura about me. Many a person passed and repassed since I was seated in a place of passage to the restroom facilities. Something about my demeanor encouraged those passersby to eschew any interaction. The recognition of being ignored heightened my negative thoughts. It further assured me that *nobody cares about me.*

My reverie of self-deprecating and self-pitying thought was interrupted when the large body belonging to Lieutenant Colonel Barnes consumed the remaining balance of the bench upon which I sat. I was surprised to have my personal pity perimeter penetrated, and with no thought one way or another, I made no acknowledgement of this trespass. For a space of time the Colonel sat silently, looming over me with his hands clasped between his legs. All at once his large left arm surrounded me, and his strong hand encompassed my shoulder, bringing me firmly up against him.

I instinctively looked up at his face, and at the instant our eyes met, he said in an uncompromising, commanding, and at the same time reassuring way, "You have a problem and you will tell me everything about it."

Was it a command, a request? I don't know. It was a statement of fact. For a moment I considered dodging the inevitable with a quip or comment, but he held me so firmly in his fatherly grip that my instinctive desire to dodge dissolved.

How do you explain it? I guess you don't. Here I was, a presumptive adult, aged twenty-three years and declared to be an Officer and a Gentleman by act of Congress under the signature of the president of the United States, buttressed against the body of another man and feebly sobbing like a child. I gave vent to my many frustrations while pouring out tears of despair. Why, why, why? How, how, how? I vented and vented, to what avail? My tears drenched his perfectly starched fatigues, and it was only when I became fatigued and silent that he began to counsel me.

"Dinan, you are a kook and constant pain in the ass." At this I tried to move away but he maintained his strong grip. "You have conjured up the idea that you are totally lacking support from those who might support you. Do you suppose that I, along with the rest of the members of the Can Tho Mess Association Board of Directors, challenged the allegations put forth by the IG, Major Nordquest because we like your smile? Do you suppose that General Eckhardt rejected a report by his inspector general in its entirety, for the first time in his long career, based on some obtuse generosity? Don't answer. I'll tell you. The people that count here believe that you are accomplishing a tremendous service, not only for the folks billeted at Eakin Compound, but also for our brothers out in the field.

"You're in every way right in thinking that you are not loved. You're not. You are respected. There are many in the command power structure who would send you off in an instant if they could find someone to replace you, but they can't. I don't know how you do what you do, but I can see that you do it.

"When you look at yourself in the mirror tomorrow morning, you'll see red eyes and dark circles and a face contorted with humiliation because of what's happened here tonight. I'll see joy and victory because you have given me absolute proof that my confidence in you is not misplaced. It is only the man who really cares that will crumble as you have. Now I want you to get up off your ass and go to bed, and in the morning, I expect you to be at your desk doing what you do best, making all our lives more livable. Will you do that?"

He released me from his bear hug and grabbing both my arms, turned me toward him awaiting a response.

I looked him directly in the eyes and stated definitively, "Sir, I'll be on the job tomorrow."

And so I was. Things looked much brighter, and I asked—and still ask—"How does the military find and nurture such incredible officers?"

Incredible enlisted personnel were also rampant, and one chap who definitely earned special mention was Staff Sergeant "Big Daddy" Hart. Big Daddy was *big*. When he was flown over to Vietnam, two passenger seats were required to accommodate his bulk. By mutual consent, his return transport was accomplished on the water rather than in the air. Big Daddy was known to be big-hearted, lighthearted, and gifted with an unusual amount of energy for a man of his or any size. His need and desire to have lots of things to do, in order to keep up with his energy, drove him into my circle as one of the enlisted men I could call upon to help out when special events required extra hands. Although these extra hands did earn extra money, the money involved was insufficient to be the real motivating factor.

Sergeant Hart invariably displayed a warm smile and an attitude of bonhomie that was as welcoming as it was disarming, yet rumor had it that you did not mess around with or cross Big Daddy. Rumors being what they are, I paid little attention to such reports and, as a consequence, found myself astonished when I chanced to observe the spectacular efficiency with which this warm-hearted Sergeant dispatched confrontation.

Inexplicably, the custom of setting up a full-sized craps table in the covered arcade between the EM club and the EM mess on payday evenings had become ritualized. Combining a full pocket of cash with a half a load of whiskey generates exuberant enthusiasm. Confronting a person possessed of these otherwise joyful circumstances with an overwhelming temptation defies any concept of prudence. The perfectly articulated plea found in the Lord's Prayer "And lead me not into temptation" was, by the setting up of this craps table, confounded completely. The abundant joy demonstrated by the winners contrasted sharply with the anguish and sorrow of the losers, and all gave vent to their emotions in what may be described as Biblical proportions.

History as well as reason informed me that these gaming nights could turn explosive. As I was responsible for both the clubs and mess halls, I made a habit of keeping a careful eye on the situation in order to head off probable conflicts that might spill over into those facilities and result in serious material and physical damage.

On one of those evenings, I was privileged to be an observer when Sergeant Hart happened by and felt the need to prevent a young soldier from inflicting more damage to his purse and person than he had already accomplished. It was perfectly clear that the soldier in question had diminished his normal acumen by giving unwholesome support to our beverage sales. It was equally clear that his proficiency in dice hurling was not in sync with his enthusiasm for that sport. Having made an assessment of the situation, Big Daddy guided the young man away from the craps table and quietly, but firmly, suggested that he might better serve his best interests by engaging in sleep rather than in stupidity.

Unhappily, our young soldier had crossed that line where sober reflection keeps us from self-destruction. Firmly, and not at all quietly, the young man expressed very negative assertions regarding Big Daddy's person and purpose. The use of the expletive-deleted device would render a record of his diatribe to the use of two words: I and you. I was intrigued to notice that all during this horrible onslaught Sergeant Hart was simply looking directly into his assailant's eyes while maintaining his usual benevolent smile. Although this young man equaled the Sergeant in height and was graced with the broad shoulders and ample chest of a natural athlete, his body displacement was unequal by more than half. At a certain point, Big Daddy had had enough. With his usual smile still firmly affixed upon his cheerful face, Big Daddy raised his big arms over his big head and in an instant brought his big hands down over the ears and onto the shoulders of his verbal assailant, plummeting him instantly to his knees.

Shock and awe are the only appropriate words to describe the facial expression of this young man as he was delivered to the ground. Like the recipient of a bolt of lightning, he was rendered helpless and speechless. Using those two great hands, Sergeant Hart then lifted him up like a twig and without the least bit of hostility told him, "Soldier, you have had your day, and tomorrow is another. You will now leave this area and go to your quarters, and you will thank me in the morning."

In the morning, he was thanked. How does the military find and nurture such incredible enlisted personnel?

THIRTY-TWO

TET AND OTHER HOLIDAYS

DAYS, weeks, and months passed with repetitious similarity interrupted only by Thanksgiving and the Christmas season. Family holidays are always tough on those far from home and hearth. For military personnel far away in a strange country, where some members of the indigenous population are actually intent on killing them, these times of comfort and joy are marked by despair and depression. In order to neutralize this negative, it is customary for mess halls and clubs to go to all possible lengths to provide festive fare and furnishings. Naturally, the Can Tho Mess Association went all out.

For Thanksgiving, along with a menu selection boasting every conceivable traditional dish and dessert, we constructed huge cornucopias for display on our service stations. These horns of plenty were stuffed and overflowing with fruits, nuts, breads, and bottles of wine. With all the goodies spilling out in seemingly endless abandon, that sense of America's bounty and blessings was clearly displayed. These displays of America's prosperity lifted many sagging spirits and caused many to reflect with true thanksgiving.

Early in December, the daily mail morphed into the daily mountain. Several parking spaces had to be relocated to provide space for sorting the heartwarming onslaught of gifts sent by loving families and friends from the States and all around the world. How the Military Postal Service was able to gear up and handle this volume for so many days is another wonder of the war, but unquestionably, we were treated to a daily on-the-ground visual of why our soldiers working in this MOS are so highly esteemed.

Christmas is a holiday that begs for an overkill of sumptuous decorating. To enhance the festive allure of the many lighted and ornamented trees in the clubs and dining rooms, we collected hundreds of empty boxes, gift-wrapped them in the glitziest of papers, and placed them under the trees with parental perfection. Each tree looked like the one you would have liked to have found as a child on Christmas morning. Abetted by a full-blown nativity scene out-of-doors, along with many more decorating highlights and a steady stream of Christmas music, the entire compound resounded with comfort and joy.

Christmas celebrations were quickly followed by our New Year. If the joy of the night before can be measured by the suffering the morning after, we can conclude that great joy abounded. The joy to be found at the enlisted club was so incredibly enthusiastic that the club was shuttered well before the midnight hour. The officers managed the occasion with less overt enthusiasm and "Auld Lang Syne" rang out at the stroke of twelve. New Year's Day was not a pretty sight in our clubs.

Next on the agenda was the traditional Vietnamese New Year celebration, Tet Nguyen Dan, commonly called Tet. 1968 was the Year of the Monkey and a natural opportunity for some scurrilous monkey business.

Although there was an excitement in the air generated by our Vietnamese friends who were contemplating the joys of serious celebration, Tuesday, January 30, 1968, was, for us, just another typical Tuesday on Eakin Compound. Following club-closing procedures, beginning at 0100 hours and concluding at the earliest possible moment thereafter, I returned to quarters and tucked myself in. I dropped off to slumber land comforted by the regular sounds of explosives usual to this celebratory evening.

In what seemed like a mere moment, my roommates were up and at the ready in a state of high agitation. Although I was much encouraged to get up and join them on some seemingly quixotic adventure, I managed to decline with such force that they left me to my comfort. My comfort was next disturbed by some unmemorable gent who sincerely advocated in favor of my immediately heading for safe harbor. We soon parted company with mutual commitments to the other's well-being. To my chagrin, Captain Owsley arrived swiftly on the heels of that well-wisher.

Captain Owsley was big—actually, huge. Standing at something around six feet, five inches and sporting a beer belly of substantial proportions, he was a man to behold. The first time I encountered his person, he had just arrived on base and was introducing himself to one of our delicate barmaids in the officers club by hoisting his belly onto the top of the bar and instructing the petite and bewildered lady to, "Fill her up." Thankfully, on that occasion the Captain was fully clothed in regulation combat fatigues.

On this occasion, the Captain presented himself in a more stripped-down version of combat readiness. Having barged into my serene, cool, and darkened quarters, Captain Owsley proceeded to brighten my surroundings and dim my tranquility by turning on all the lights within my lair and arousing me from a well-earned sleep.

"DAANAN, DAANAN," he bellowed, while standing over my prostrate body.

I came somewhat into wakeful consciousness only to behold this mammoth looming over me adorned in combat boots, regulation issue olive drab boxer shorts, combat vest, and helmet, supported by multiple ammunition necklaces about his shoulders in concert with the M16 gripped threateningly in his oversized left hand.

When he perceived that I recognized him as the threatening behemoth that he was, he continued, "Daanan, you will get your young lieutenant ass up from that bed, grab your weapon, and report to your bunker in five minutes, or I will personally drag your ass out and deliver you as is."

Without waiting for a response, the Captain spun away and exited with purposeful haste. There was no question in my mind that he meant what he said and could easily accomplish his purpose. Armed with this certain knowledge, I resigned myself to the inevitable and scraped my body out of my bunk. I put on my regulation fatigues, strapped my .45 automatic about my waist, slipped into my flip-flops, and topped myself off with my snappy baseball cap. Thus prepared, I then poured a serious scotch on the rocks, lit a Parliament cigarette, and quit my quarters.

Although it was well after three o'clock in the morning and well before sunrise, the sky was ablaze, and the visibility equaled that found under the noonday sun. The exhilarating sounds of fireworks exceeded any Fourth of July I have ever experienced. It was overwhelming: the tracer bullets pouring out of our airborne armament in flaring red, the screaming engines of multiple aircraft, the ratta-tat-tat of machine guns firing with complete abandon, the ping-ping-ping of sniper bullets all about. I turned left and strolled toward my assigned bunker.

Although there was a host of certified unfriendlies without our compound intent on delivering bodily harm to my person, I understood that their designs had more to do with combat enthusiasm than any personal animosity. As I neared the bunker, I was unfortunately confronted by a decidedly *personal* unfriendly within who definitely harbored intense personal animosity toward me.

Newly-minted Lieutenant Colonel Archibald I. Schelling had of recent date hosted a party to commemorate his elevation from the rank of major. He had insisted upon a lavish celebration with the most and best of everything. This we delivered; this he received. He was filled with exceeding great joy until he received his bill. At that moment his joy was jaundiced, and challenging every aspect of the bill, Schelling shelled out more hours in protest than he had in planning. In the end, following an unprecedented detailed review, Schelling was ordered to shell out the cash. In the process Colonel Schelling and I became combatants. I won. Whoever suggested that "to the victor, go the spoils" was definitely referring to inter- rather than intra-military conflicts. Rather than spoil filling my pockets, my gain was that of an enemy posted in a high place and intent on spoiling my personal and my professional aspirations.

The Colonel, although designated an Officer and a Gentleman by Act of Congress, was unable to deport himself in a gentlemanly manner and accept with grace the fact that he had lost. Not only had he lost, but he had lost in battle with a lowly lieutenant. So violent was his reaction to this esteem-depleting position that he determined to exact revenge upon our little officers club. Having complied with the order that he must pay his bill in full, Colonel Schelling went on a rampage to exact some sort of Pyrrhic victory by destroying club property and provisions to any amount that would exceed his expenditure. His visits became so costly and destructive that I was compelled to make a formal request that he be barred from the facility. This request resulted in the Colonel receiving orders stipulating that he was not permitted to enter the officers club.

As a general rule, junior officers do not admonish or interfere in the activities of senior officers. It was manifest that the very same lowly lieutenant who had initiated the Colonel's billing had likewise initiated his being barred from the club. It is therefore not surprising that Lieutenant Colonel Schelling did not look upon Second Lieutenant Dinan with warmth or high regard.

So it was that when I approached the Colonel on that spectacular occasion, I was regarded with unconcealed hostility. We had a brief, unpleasant, and unproductive exchange.

Colonel Schelling began with a question, the answer to which he had already observed, "What's that thing in your hand, Lieutenant?"

Having things in both hands and not knowing which he had focused on, I responded, "In my right hand I hold a scotch on the rocks. In my left hand I hold a cigarette. A Parliament cigarette, Sir."

"What kind of stupid soldier are you? Were you sleeping as usual when the lesson was given about how the glow from a lighted cigarette can draw enemy fire?"

I passed over the first question and addressed the second, "I recall being awake during that lesson, and the draw of enemy fire to such a glow was described as something likely to occur when one is surrounded by darkness. This place is lit up like high noon, Sir."

"Well, where's your weapon, soldier?"

I tapped my right hip and replied, "Right here. Sir."

"What the hell sort of supposed weapon is that?"

"This is a .45 automatic. Sir."

"Why do you have a .45 automatic, soldier?"

"When I was processed through in Saigon they had run out of .38's, so this was the smallest thing they had left. Sir."

At this, he glared at me and shook his head in obvious disbelief. He then continued his interrogation, "Why aren't you wearing your helmet, soldier?"

The soldier thing was getting a little boring, but as a senior officer he was entitled to belittle me in this manner. I figured that he was doing this out of personal animosity, and really, there was nothing I could do about it anyway, so I let it go.

He seemed genuinely shocked when I told him, "I wasn't issued a helmet. Sir."

"What do you mean, you weren't issued a helmet, soldier?"

"When I was being processed they asked me if I wanted one and I said no. Those things always give me a pain in the neck. Sir."

"Soldier, you give me a pain in the neck, and I hope you get your head blown off. Now, put out that cigarette and get into the bunker."

No question there, so I said nothing and headed for the bunker. When I reached the entryway, the violent buzzing of mosquitoes all but drowned out the sounds of war blasting all about me. I turned and faced the Colonel, who had followed right behind me, and said, "I can't go in there; it's too dangerous. I would be eaten alive. Sir."

"I forbid you to return to your quarters. That's an order, soldier."

I looked about and suggested, "What if I just take a seat over on those steps. Sir?"

Stealing a line from *Gone with the Wind,* he responded, "Quite frankly, I don't give a damn. Go right ahead soldier, and I hope you get yourself killed."

No question there either, so I sat myself down and enjoyed a spectacular display of military fireworks. It was a feast for the eyes and ears. At the same time, the Colonel went into the bunker and joined the feasted upon.

In reflecting on that night, I wonder how it was that I experienced no sense of danger. I suppose I was simply too intrigued and too naive to be struck with the very real and apparent danger, even though I knew that Eakin Compound was a prime target. Everyone in the Can Tho area, friend and foe alike, knew that the commanding general and his entire senior staff, commissioned and non-commissioned, was quartered in this small complex. Killing off this bunch of brain power would be like removing the head from an otherwise strong and capable body. The VC knew this. If they could wreak destruction on Eakin Compound, they could destroy the American advisory effort in the Fourth Corps. So we were surrounded and taking fire from small arms, rockets, and mortars. The newly arrived Cobra helicopter gunships hastened to our defense while Puff the Magic Dragon flew overhead dropping flares and firing heavily upon exposed enemy positions. Without the combined and coordinated actions of these airborne arsenals, we would have been overwhelmed and overrun. Talk about the cavalry coming to the rescue! In any event, I enjoyed the show and disobeyed my orders at one point by returning to my quarters to refresh my drink.

As the sun began to rise up, the racket began to quiet down, and soon those who had spent the night huddled in the bunkers began to emerge. It was a ghastly sight to behold. Bitten, bruised, and exhausted, our fearful warriors appeared one by one into the daylight. As they showered and shaved, I inspected the mess halls. Finding everything up and running, I retreated to my refrigerator, popped open a bottle of champagne, and drank a toast to the Year of the Monkey, the new day, and "Yesterday."

It was a new year and a new day. Everything that was had gone away. Nothing would be the same as yesterday.

> Yesterday,
> War was such an easy game to play.
> The VC seemed so far away,
> Oh, I believe in yesterday.

At sunlight, when most of us were trying to figure out how to enter into this new world of real war, Captain Spain entered the cockpit of his spotter plane. His mission was to observe and report on various enemy movements, buildups, strongholds, and whatever else might be of military use. At this early hour there was confusion in the ranks. Although self-defense was always allowable, and that defense had been vigorously and effectively rendered throughout the night, the Central Command in Saigon had not communicated an end to the Tet truce. With that pact in place, any aggressive action on our part not provoked by enemy fire was strictly forbidden.

As Captain Spain flew his Bird Dog at low altitude up the Bassac River, he clearly observed a convoy of sampans floating downriver toward Can Tho. He noted that the boats were drawing water over their sides, a sure sign that they were carrying too much weight. He noted that the people on board waved Viet Cong flags and welcomed his inspection with vigorous waving. He further noted that the waving hands were clenched into fists save for the middle fingers that were fully extended. He wanted to return this greeting by unleashing the rockets he had attached under the wings of his craft as a suitable response to this unsuitable welcome. Alas, he was bound by the truce.

Continuing his reconnaissance up the Bassac, he maintained full radio contact and made regular reports on his location and observations, all this while his sense of humiliation, helplessness, and anger caused by his encounter with the sampans continued to grow. Captain Spain was not a happy flyer during this interlude. In time-honored tradition, he beseeched the Almighty to do something, anything, anything at all, to rescue him in his hour of need. What he needed desperately was the go-ahead to return the greetings extended by the flag-waving VC antagonists.

In all likelihood the Almighty had nothing to do with it, but his prayer was answered, and it was the answer he wanted.

As the sun continued to rise in the east, these soothing, electrifying, and energizing words poured out of his earphones and into his welcoming ears: "The Tet truce has been called off. Repeat, the Tet truce has been called off."

This was truly gospel to this pilot. Could this be true? Captain Spain was already forming his plan of action as he turned his craft about and headed downriver.

Ever cautious and professional, he radioed in to central control at Can Tho Airfield, "This is Shotgun. Can you confirm the message I received about the truce?"

This happy reply was given, "Shotgun, this is central control. If the message you received said that the truce has been called off, that's a Roger. Repeat, message confirmed. The Tet truce has been officially called off by Command Saigon. Shotgun, do you copy?"

"This is Shotgun. I copy. Over and out."

Exercising almost inhuman personal restraint, Captain Spain maintained a very slow observation airspeed. He had no way of knowing whether or not the enemy below had received a similar transmission. He figured that there was a good chance they hadn't. Just in case, he would be very alert and very careful.

As he closed in on the convoy, lazily moving along with the current downriver toward Can Tho, the joyful crews on those boats once again welcomed his observation with waving flags and fingers. This time he waved back. They loved it: American flyboy sucker, drop dead.

Passing over low and slow, his observations of their every movement convinced him that although the sun was shining, they were in the dark. When he cleared their airspace he looked back and enjoyed watching their demonstrations of defiance and glee. When he passed a bend in the river that put him out of sight, he turned upriver and engaged his rockets. Combat ready, he reapproached. Slowly, slowly, at low altitude he closed the space between them. He could count their teeth as they gave him their most horrid and joyful smiles.

Taking careful aim, Captain Spain unleashed his rockets in rapid succession. Every boat in the convoy suffered a hit. Every boat sank like a stone. There was no casualty count, but the Captain observed many crew members who had lost their smiles desperately swarming toward the riverbanks to save their lives.

This happy hero radioed in his report and resumed his mission.

When it became feasible, navy divers inspected the sunken vessels and confirmed Captain Spain's suspicions. The convoy was packed to the max with heavy artillery and crates of ammunition. Had these goodies found their destination, certain death and destruction were destined to decimate Can Tho.

Military payment certificates (MPC) for 5, 10, 25, and 50 cents measured 5.5 x 11cm. MPC worth 1, 10, and 20 dollars were significantly larger in size, similar to United States paper currency. They were all quite colorful; this 5 cent MPC features lavender designs on a turquoise blue ground.

Open air market

Lieutenant Dinan and Sergeant Cole

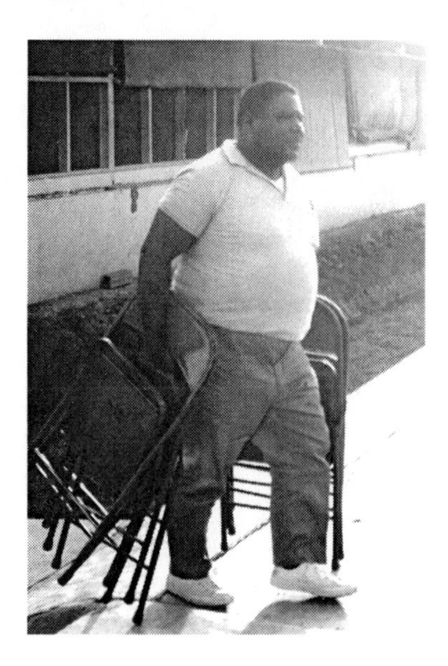

Big Daddy Hart

Thanksgiving cornucopia in the officers mess

Christmas tree in the enlisted mens club

Christmas crèche in front of compound headquarters

Amazing Christmas mountains of mail

January 1968—relaxing in C-5

BIRD DOG

The Cessna O-1 spotter plane, a two seat (one behind the other), fixed wing aircraft, was used in Vietnam for aerial reconnaissance. Its main function was locating the enemy and calling in ground troops and/or attack aircraft when needed.

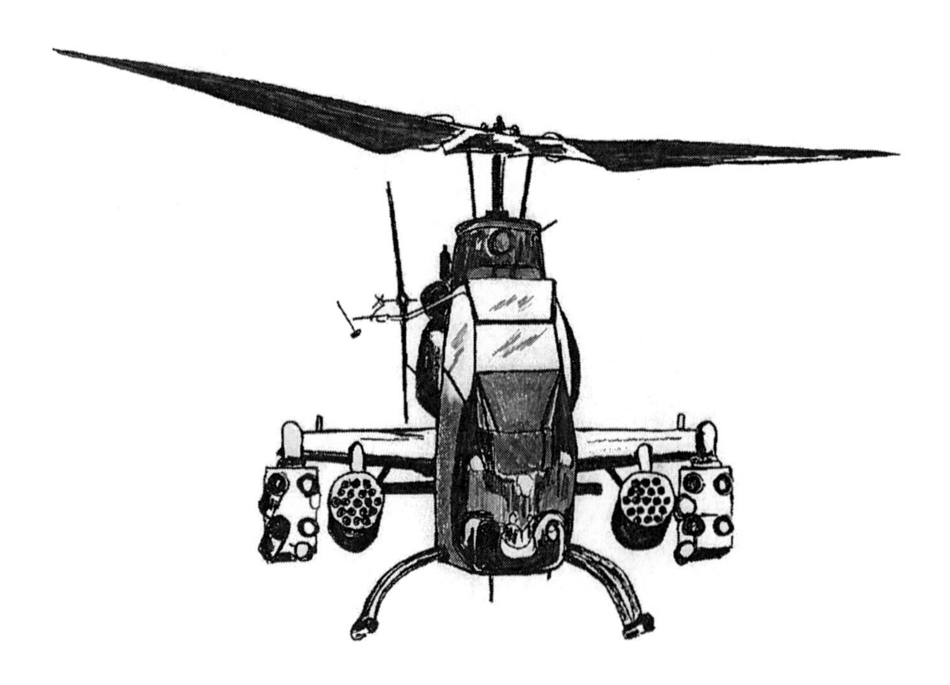

COBRA

The Bell AH-1 Cobra, a two-bladed, narrow-bodied, single engine attack helicopter armed with rockets and machine guns, was introduced in 1967. Its superior maneuverability was astonishing and vital to the war effort. It was also known as the HueyCobra.

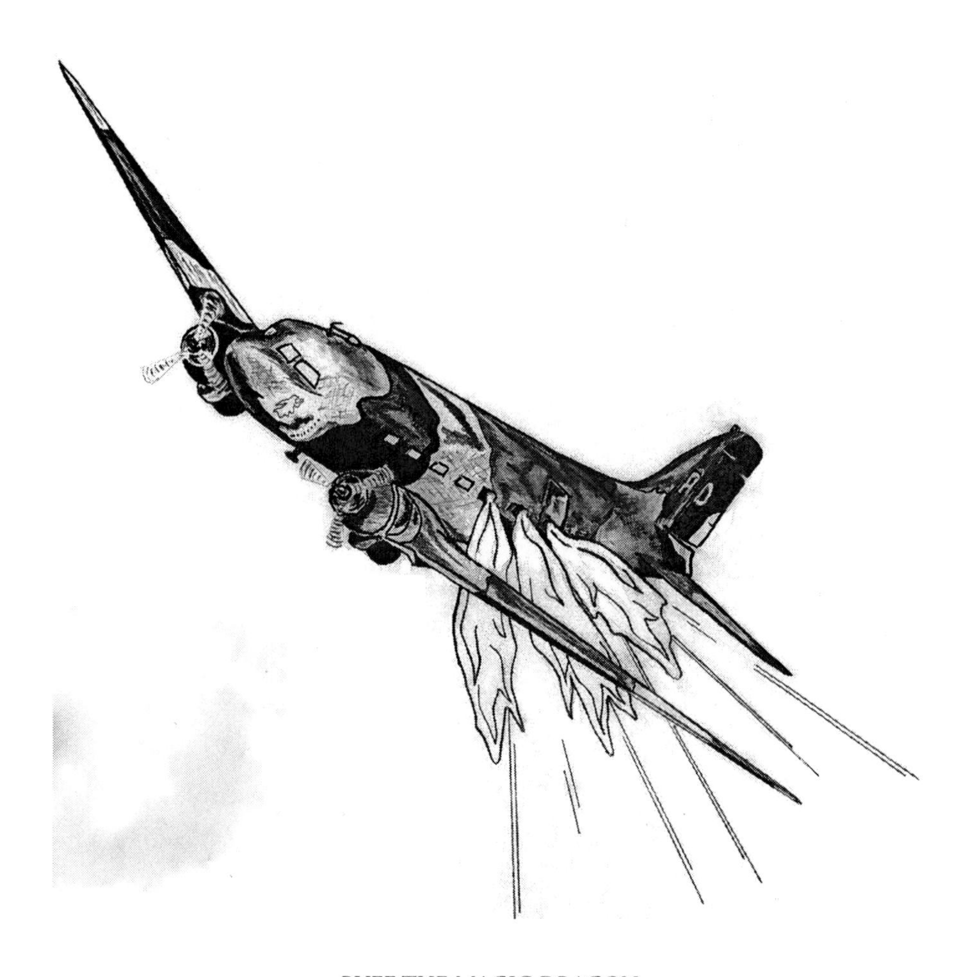

PUFF THE MAGIC DRAGON

The Douglas AC-47, a gunship also called Spooky, was developed by the U.S. Air Force during the Vietnam War. Armed with three portside mini-guns firing 18,000 rounds per minute plus flares to highlight enemy positions, its very long wingspan enabled it to fly, almost glide, slowly enough to accurately target the enemy without endangering those under siege. It circled counterclockwise and was invaluable in suppressing enemy ground attacks.

THIRTY-THREE

PAPER PROBLEMS

EVERYTHING was sort of the same—but at the same time sort of different. In the first few hectic days after Tet, we were bereft of our Vietnamese workers. The entire American contingent had to make their own beds and shine their own shoes. The laundry service was suspended. For the first time ever, the lower-ranking enlisted personnel were subjected to work details and kitchen duty. Sergeant Cole and I had to scramble to find willing replacements for our club workers. The snack bar was shuttered, and no one needed piasters. There was that sense of urgency and mission. Real combat. Real war.

From out of nowhere there appeared an Armored Personnel Carrier. This splendid vehicle transported General Eckhardt along with his chosen minions between Eakin Compound and MACV IV Headquarters. The adjacent soccer field was transformed into a helicopter facility, and important personages were whisked in and out with alarming regularity. Incoming supplies were at a standstill. All transportation was dedicated to combat needs. We were told to make do. However, there is do and then there is doo-doo.

The least celebrated and glorified of the United States Army branches is unquestionably the Quartermaster Corps. The flamboyance and derring-do that so excite our passions are the stuff and substance of the units charged with engaging the enemy. The Green Berets, Navy Seals, Paratroopers, Rangers, Fighter Pilots, Marines, the Combat Infantry—these are our protectors, our fighters, our celebrated and our well-deserving, well-sung heroes. Yet it is often and truly noted that an army moves on its stomach. Filing those stomachs is only one of the many support activities accomplished by the Quartermaster Corps.

We celebrate our combat heroes because of the terror they are forced to face. If you would not be terrified by the mission of supplying more than half a million soldiers, thousands of miles from home, with every possible need and comfort, 24 hours a day, 365 days a year, you may not understand this particular manifestation of terror. To fulfill this mission, the Quartermaster Corps has developed and refined procurement and distribution procedures, guidelines, factors, and policies that deserve wonder-of-the-world recognition. The seemingly most insignificant need has been scrutinized, evaluated, and enshrined in voluminous operation manuals. Thousands upon thousands of items from auto parts to zippers are identified, codified, and supplied. Included in this encyclopedic listing is a very necessary, basic requirement for proper human hygiene that we have come to take for granted: toilet tissue.

If you were responsible for ordering toilet tissue for Eakin Compound, what would you do? Mother Military has it figured out for you. You consult your manual. In your manual you will find a factor. The factor is to be multiplied by the number of women and the number of men in residence. An all but absolutely precise differential of use by the different sexes has been carefully determined. Thus, through the application of simple counting and multiplication, you become a procurement genius. As with most military guidelines, there is one caveat here. The person in charge must assess actual ground conditions and adjust accordingly.

In the immediate aftermath of the Tet uprising, our compound supply officer, Lieutenant Dewey (you remember him, the chap who went out of his way to order chow line feeding trays on my behalf) went to his supply cupboard and discovered that he was about to run out of toilet tissue. This was very bad indeed. Had I run out of mai tai mix, Coon Skin may have skinned me, but this would be gentle treatment compared to the pain inflicted by a major general subjected to the personal indignities inherent to the lack of T.P.

Evidently Lieutenant Dewey had failed to make a ground assessment. With all the factoring and figuring, the fine-tuning regarding the toilet tissue factor in this conflict fell by the wayside. What had been overlooked was the unhappy side effect of the required once-a-week malaria pill. Admittedly, this was a time long before the now ubiquitous warning label that is attached to almost every ingestible. Notification of the nasties associated with this necessity would have been helpful. However, the unnoted, notable nasty of this medicine was a regular and recurring weekly assault on the gastrointestinal apparatus, and this fact severely altered usual operational efficiency. Bypassing gross description, it will suffice to report that the use of this paper product far exceeded official predictions. Thus, we had a dilemma.

Those operationally involved in this dilemma quickly resigned themselves to three pertinent realities. First and foremost, we could not run out of toilet paper. Second, the only supply source was situated in Saigon. Finally, no air transport would be allocated for any noncombat related purpose. Bracketed by these agreed upon realities, a plan of action began to fall into place. The basic operational goals were straightforward and simple:

- Get to Saigon
- Secure toilet tissue
- Return to Eakin Compound

How does one find their way from Can Tho to Saigon when they find themselves with no access to the waterways, restricted from air travel, and facing countryside between those two locations that is crawling with killers?

It is little remembered and rarely mentioned, but even at the height of hostilities, highways between major cities remained open for business. It was a question of power, not political or military power, it was *petrol* power. Although there was a considerable separation of thought concerning political power, the populous shared equally in their commitment to, and need for, the power provided by petroleum. This shared need resulted in a boon for the international oil companies operating throughout South Vietnam. The fuel power they provided was protected, and the roads needed to haul their power to the people, of whatever political bent or stripe, remained open and in reasonable repair. Although large, lumbering, and easy to mark, no petroleum transport was ever assaulted.

This anomaly provided an opportunity for a reasonably sound strategic overland plan. This mission improbable was dubbed by our enlisted comedian contingency as Operation Ass Wipe.

Once it was agreed that OAW would be executed using ground transportation, the obvious question concerning vehicle selection was placed on the table. The use of military transport of any kind was unanimously dismissed. What then were the alternatives? Ox carts and such were offered as possibilities, but quickly rejected on grounds of speed and payload. One of the planners recalled seeing a little yellow school bus belonging to the local Roman Catholic school puttering about the city and recommended its use on this noble adventure. This quixotic thought was first greeted with much derision and scorn. In the end, after much deliberation, Don Quixote was called upon to slay the windmills aligned against OAW.

Under veiled circumstances and agreements, the yellow school bus soon found its way onto Eakin Compound. Most of the seats were removed, and the interior was fitted with metal plates. The old tires were replaced with brand-new–military-issue truck tires. The engine was fine-tuned by mechanics brought over from Can Tho Airfield. Filled and fitted, the innocent little yellow school bus was now mission and combat ready.

At sunrise, after Sergeant Erickson and his staff had packed the bus with provisions, the chaplain gathered the expeditionary force by the side of the bus and offered prayers of protection and a grace. A sizable group of onlookers joined in the solemnity of the occasion and added their encouragement and prayers. The select group of volunteers, packing enough armaments to defend the Alamo or hold off an enemy battalion, mounted up and moved out. To a man, we feared the worst and hoped for the best.

Two days later, the little bus puttered back into the compound heavy laden with toilet tissue. Universal exclamations of gratitude and joy reverberated over the entire camp. Operation Ass Wipe was declared an unqualified success. Not an enemy seen, not a shot fired. The roads had been empty and open. The sun had shone.

The post-action report stated: "It was like a Sunday drive through the park." The power of petrol had prevailed.

THIRTY-FOUR

HAPPY BIRTHDAY TO ME

FRIDAY, February 2, 1968, was my birthday. Of course, all February 2nds share this annual distinction with Groundhog's Day. What has kept this particular birthday most memorable for me was not the day, but the night, when, as officer of the day, I was on duty all night.

I had done this duty a number of times before, and for the most part, it had simply been boring and inconvenient. Of course, that was yesterday. This was an entirely new game to play. Real spice was added to the game by the addition of a squawk box or listening device. This thing spewed forth an unrelenting, steady stream of barely intelligible babble. In concert with this was the addition of a large red button suitably designated "the panic button." This thing practically glared at you, while offering up the silent challenge, *Push me if you dare.*

The challenge in this game was for the OD to decipher the squawks coming out of the box and, if convinced that real and present danger was imminent for Eakin Compound, go into a panic mode and push the button. In the event the button was pushed, glaring lights and screaming sirens would be activated. This single act would interrupt the sound slumber being enjoyed by the general, the colonels, the lieutenant colonels, the majors, the sergeant major of IV Corps, the first sergeant, and a host of master sergeants. Need I mention the less exalted comrades at arms?

In this game no score was kept. Either you won or you lost. Losing was far too ominous an option, no matter which of the two losing scenarios concluded your game. In the first scenario, you panic prematurely, unnecessarily disturbing the repose of those who are not to be disturbed. In the remaining scenario, your fear of the consequences of crying wolf causes you to remain silent, when as a good shepherd you should be sounding every alarm available in an effort to protect your flock. In either case, should you happen to remain alive, you might consider the alternate option preferable.

Winning was also a two-scenario adventure. In the first scenario you won if by doing nothing, you did the right thing. In the alternate scenario, the enemy advance is so brightly illuminated that you courageously imitate Paul Revere, sound the alarm, and ride off into midnight as a national hero.

Of course this was not a game. The responsibility of securing the very lives of more than three hundred comrades was more frightening than anything I could have imagined. I was extremely fortunate to have Master Sergeant Serge Kowalski assigned as my charge of quarters for that night's specific tour of duty.

I arrived at Compound Headquarters before Sergeant Kowalski. Upon his arrival, I was introduced to him as his OD.

Unflinchingly, he summed up the situation by announcing, "Lieutenant Dinan, I vil be vit you. Ever ting vil be OK."

Lean, mean, and a military machine, Sergeant Kowalski was every inch a real-life lifer. He had served with distinction in the Polish Army prior to his enlistment in ours. Standing ramrod straight before me, he projected a combination of self-assurance and deference to rank the likes of which I had never encountered. There apparently remained that something in the European military mentality that rewarded the officer ranks with an automatic deference not to be found in American-trained enlisted personnel.

I found myself happy and humbled by my good fortune: happy to have a seasoned, experienced professional I could lean and depend on; humbled by the realization that although I was the designated leader, my subordinate was far better prepared to take charge. Before the sun set, we made our first perimeter inspection. Walking parallel to the outer defense line around the entire compound, we ensured that all the fencing was secure and that the guard towers were properly manned and equipped.

While the sun still shone, everything was calm and quiet. Under these conditions it was a rather pleasant way to pass some time. As the sun began to disappear, the sounds of war reappeared all about us. A symphony of hostility filled our ears with an unrelenting repetition of boom, boom, boom, ratta-tat-tat, ping, ping, ping. Trying to ignore this unwelcome music, we concentrated our attention on the sounds coming from our little squawk box. During our entire tour of duty, we never were able to make out even one complete, intelligible sentence. Most of our conversation throughout that night was confined to: What did they say? What does that mean? I don't know. I have no idea. What we heard was:

- Mumble, grumble, garble, enemy troops crossing the Bassac at garble, mumble squawk.
- Can Tho mumble, garble, garble squawk, under heavy, squawk squawk grumble.
- Large enemy force observed, grumble squawk, garble garble mumble.
- Squawk garble mumble infiltration, garble garble forces squawk, garble fire.

And all the while, all about us, boom, boom, boom, ratta-tat-tat, ping, ping, ping. And all the while, all but shouting, the button beckoned, pleading to be pushed.

At midnight our first scheduled relief arrived. Some relief. The arrival of these temporary replacements signaled that it was time for us to abandon the relative safety of our post in Compound Headquarters and once again walk the perimeter. Covering our heads and bearing our arms, we entered into the darkness and began our circumnavigation, walking beside the long wall of the soccer field. My head gear was my baseball cap, and my weapon was my .45.

Sergeant Kowalski, more experienced and more sensible, covered his head with a standard army steel helmet and toted an M16 rifle.

The wall was thick and stood shoulder high, providing what I judged to be sufficient protection. The more experienced and knowledgeable Sergeant concluded otherwise when he judged that the ping, pinging sounds all around us were getting too close, upfront, and personal. Assuming the prone position, he proceeded to advance along with me by executing an excellent low crawl. I was vigorously encouraged to follow suit, but I declined. It wasn't a matter of fearlessness. It was the result of conflict resolution between two conflicting fears.

The fear I knew very well revolved around the bodily harm I incurred when participating in the sport of low crawling. I had been forced to participate in this activity on many occasions during my military training. The results of my noble attempts to engage in this mode of locomotion evidenced, on all occasions, the fact that I was not well formed for this purpose. At the completion of such exercises, I would rise up from God's good earth and look down upon a torn and bloodied body—mine. The shreds and tears inflicted upon the strong cloth that covered my delicate skin paled in comparison to the deep lacerations inflicted on and about my knees and elbows. The ensuing pain was memorable and reminded me that this was not the way for me to travel.

The fear I knew only in the abstract centered on the possibility that I might find myself in between the discharge of a bullet and some random point of impact. After all, who would be so base as to actually shoot *at* me? This was not necessarily clear thinking, but providence prevailed, and I persevered unharmed. In the meanwhile, the Sergeant and I presented a tableau very much like a man walking his dog or pet alligator.

Our inspection tour successfully completed, we returned to our post. Bidding our relief unit sweet dreams, we resumed our listening stations. I had visions of being blessed with a more acute listening capacity based upon a new appreciation of our vulnerability. This hoped-for capacity never materialized, and I remained dumbfounded trying to decipher the messages that were contained in the squawk, mumble, grumble, and garble interspersed with snippets of intelligible words.

We welcomed our 0300 relief with resignation and trepidation. In accordance with the finest military tradition, we reentered the field of combat and successfully completed our final perimeter inspection.

Things had quieted down by this hour, and our final inspection tour was somewhat less frightening than our midnight adventure. Yet gremlins, goblins, and other underworld creatures with malignant design still seemed to move in every shadow. The thing that really struck me during this inspection was our compound's vulnerability. It was as though I was observing our defenses for the very first time. I imagined that if the enemy saw what I saw, we would be overrun and obliterated. I had no in-depth training in perimeter defenses, and this specialized military function was worlds away from my primary MOS. Not wishing to appear completely stupid, I said nothing. I was happy enough to return to headquarters in one piece, listen, observe, and, with great joy, welcome the break of day.

Promptly at 0600 hours, Sergeant Kowalski and I were properly relieved of our post. One of the general orders instigated by General Eisenhower clearly states, "I will quit my post only when properly relieved." Unwritten military tradition allows that an enlisted person or non-commissioned officer who has endured an all-night posting is entitled to the comfort of bunk time following such a tour of duty. The commissioned officer, however, is expected to buck up and proceed with the new day's activities as though possessed of a full night's sleep. Accordingly, the charge of quarters charged off to bed, and this officer of the day began his daily duties.

Supercharged by the combined joys of being alive and of having resisted the many impulses to panic and press the button, I entered into this new day and new year in my life—I was now officially twenty-four—with fortitude and energy.

By mid-afternoon, fortitude and energy had morphed into fatigue and emptiness. My commissioned status notwithstanding, I sought and secured the comfort of my bed.

By the time I got up, the sun had gone down, and the orchestras of the armed once again filled the air with the symphony of war. A veritable 1812 overture: da-da-dada-dada-da, boom, boom, da-da-dada-dada-da, boom, boom. Under the circumstances, this music was anything but music to my ears. All alone in the darkness, I contemplated my best course of immediate action. Finding myself too groggy and grungy for any clear thinking, I headed to the showers for purification and renewal.

Standing under the downpour, my mind began to wander. Those hollow sounds peculiar to a large enclosed space devoid of carpeting, upholstery, or human company started to take on an eerie quality. In the dim light of the shower stall, I listened to the echoes of the cascading water bouncing off the bare walls, accompanied by an ominous repetition of boom, boom, boom, ratta-tat-tat, ping, ping, ping. In its wandering, my mind found and settled in on visions of our vulnerability. Every area in our perimeter defense that I had judged during my inspection to be soft now came into hard focus. I imagined hordes of Viet Cong bursting through these pitiful defense lines. I imagined them to be armed to the teeth. I imagined them to be filled with blood lust. I imagined them assembling outside the door of my solemn and solitary echo chamber. I thought, *What shall I do when they rampage my sanctuary?* I took stock of my situation and thought, *Here I am, completely alone, completely naked, armed with a washcloth; what can I do?*

A terrifying fear gripped me completely. I looked all about, but there was no escape. As I looked down upon my exposed and defenseless nakedness, tears of anguish and sorrow welled up in my frightened eyes. My heart felt as though a cold steel hand had reached into my chest and was now squeezing it. I could feel the blood pumping into my head. I could feel the sweat pouring out of my body. I felt hot and cold at the same time. I felt as though I would fall, so I sat on the unwelcoming floor and bowed my head between my shaking knees.

Scared to death! How long did I sit with pounding heart and head? As my senses returned, I understood, for the first time, the real meaning of that oft-used and mostly trivialized declaration. I understood for the first time why unnumbered soldiers had taken their own lives for fear of death. Scared to death! I was truly scared to death. The absoluteness of the pain and panic is beyond telling.

I scraped myself up off the floor, shut off the shower, wrapped myself in a towel, and quit that chamber of horrors in desperation. Once returned to my quarters, I poured a stiff brandy and huddled on our sofa. My first competent thought was *I cannot allow myself to get trapped into that sort of thinking ever again.* It was simply too frightening. I wasn't sure if I could survive a repeat of that experience, and I didn't want to find out.

When I felt sufficiently calmed, I dressed and made a beeline for the club in search of companionship. I must have looked more the worse for wear than I realized because a number of fellow officers asked if I was okay. I satisfied their kind questioning with tales of my OD duty. It is possible, even probable, that I would have benefited myself as well as my kind questioners by sharing a full account of my scared-to-death experience. I couldn't do it and, for many years, I never did. And although I never let myself get caught in that trap again, it is a memory that I can never completely shake.

THIRTY-FIVE

SEARCH AND WHAT?

AFTER a little more than a week, combat conditions in our immediate area quieted down, and our Vietnamese staff was thankfully allowed to return to work. At the same time, Eakin Compound hosted hundreds of real American combat soldiers for a few days. Without the assistance of our Vietnamese friends, we would not have been able to provide the hospitality that these fine young men so richly deserved. As it happened, we were able to lay out the red carpet, and I particularly enjoyed the sight of these men in full combat regalia roaming about the compound with rifles in one hand and ice cream cones in the other.

One of the cone carriers I recognized from my high schools days at Evander Childs High School in the Bronx. Quentin Joseph DeSantis had been on the football team and even in combat gear, he looked like a linebacker. Sergeant DeSantis, United States Army, 4th Infantry Division, filled me in on the ridiculous and unhappy happening that thrust his division unnecessarily into harm's way.

The enemy combat forces that had been driven out of the immediate Can Tho vicinity were holed up and regrouping in the land mass between Can Tho Airfield and Binh Thuy Air Force Base. The highly acclaimed and much decorated Vietnamese Ranger Battalion, charged with performing a search-and-destroy mission to rout these unfriendly forces, had been specially flown in by U.S. helicopters. Although all actual ground combat operations in the

Fourth Corps were exclusively reserved for Vietnamese forces, this operation was judged to be so important that the U.S. 4th Infantry Division was brought in as backup.

The idea was that the U.S. force would assemble a parallel assault line some distance behind the Ranger Battalion and advance at whatever pace was established by the forward line. This way, if our allies got into a bad spot, we would be able to move in quickly and come to their aid. The operation was designed so that any first contact with the enemy would be made by the Vietnamese forces. It did not work out that way.

While maintaining constant communication, the U.S. 4th Infantry advanced in sync with the Vietnamese Rangers. At every moment, the Americans were ready to close in rapidly and provide backup aid and firepower in the event that a successful search effort resulted in a destroy operation beyond our allies' capability. On the ground, out in front of our forces, the Vietnamese Ranger Battalion encountered numerous successes in the search aspect of their mission. However, rather than engaging the enemy and activating the destroy part of the mission, this highly decorated unit performed a maneuver that Sergeant DeSantis referred to as "search and avoid." That is, when the enemy was discovered, the Rangers carefully avoided any contact and, using great stealth, bypassed them undetected.

Thus it was that the enemy's significant emplacements remained undisturbed as the frontline advanced beyond their positions. Moving forward over what was reckoned to be safe terrain, the backup assault line was alerted to the enemy's presence by a fusillade of fire rather than a phone call. Safely beyond harm's way, our friends, the Vietnamese Rangers, continued on their walk through the countryside. With friends like these, who needs enemies?

The 4th Infantry Division ultimately overwhelmed and completely destroyed the opposition. This victory was filled with defeat because we had been effectively walked into a trap by the forces we had come to reinforce.

Following this successful cleansing of the countryside, the active war moved far away from the Can Tho area. We still suffered occasional rocket and mortar attacks on the compound, but in general, things were more peaceful than they had been prior to the Tet Offensive. Even so, the realization of our vulnerability had been indelibly imprinted upon the residents of Eakin Compound. Many good men spent their last months in-country sleeping in the bunkers rather than in their beds. I opted for my cool, comfortable bed, but now I truly understood and respected their caution.

As things returned to a quieter state, General Eckhardt resumed his practice of personally overseeing the awarding of military honors. I was summoned by the General's aide-de-camp to report to the General's office at exactly 1530 hours. When I arrived, I joined Captain Giardino and Staff Sergeant Froman. Each of us had been previously informed of his own award, but none of us knew what was awaiting the others.

When General Eckhardt was ready to receive us, we were ushered into his office by the Sergeant Major of IV Corps, Sergeant Marcille. After the obligatory saluting-in, where we in unison came to attention before the General and declared, "Reporting as ordered, Sir," General Eckhardt gave us all a warm welcome and assured us that this duty was the most important and joyful of all the duties he had as a Commanding General.

Military awards are officially bestowed in accordance with the standing of the award rather than the standing of the awardees. The first medal to be awarded was the Silver Star. Sergeant Froman was called forward to receive this coveted honor. Sergeant Froman had surely distinguished himself during his year of service in Vietnam. Arriving as a Private First Class (E-3), he had risen through the ranks to the distinguished rank of Staff Sergeant (E-6) in record-breaking time. During his meteoric advancement this fine young soldier had, with unique courage and dedication, carried out his *only* assigned task: driving the colonel serving in the position of second in command back and forth between Eakin Compound and Command Headquarters. Distinguished service indeed.

The second award was the Bronze Star. The designated recipient was Captain Giardino. Concerning such medals, the military has established an Order of Precedence. In this order, the Bronze Star is ranked four points below the Silver Star. Between these two celestial awards are found the Legion of Merit, the Distinguished Flying Cross, and the Soldier's Medal. Clearly, the Silver Star shines more brightly than its Bronze associate.

Captain Giardino had arrived at Eakin Compound about six months earlier, following a stint in a military hospital. His active engagement in that facility was that of a patient. Prior to this deployment, the Captain had served as the leader of one of our advisory teams in the field. During a fierce engagement with enemy forces, Captain Giardino suffered severe wounds that resulted in the award of a Purple Heart, accompanied by numerous letters of commendation celebrating his valor and courage. Upon release from the hospital, he declined the offer of redeployment to the United States and was subsequently assigned as the officer-in-charge of security for Eakin Compound. In this capacity he reorganized much of the security structure and presented to the unfriendlies a Maginot line that deterred crossing. Further, he increased troop morale and introduced training exercises that brought his unit up to combat readiness at the most critical time.

Like the young man who had just been awarded the Silver Star, this Purple Heart recipient's service in Vietnam had in every way been meritorious. Unlike the Sergeant, however, the Captain had fought in direct combat with the enemy. Unlike the Sergeant, the Captain had exercised military initiative, craft, and leadership in the performance of his several duty assignments. Unlike the Sergeant's chauffeuring, the Captain's personal efforts unquestionably saved American lives, as attested to in his Purple Heart citation, and very likely protected many more through his efforts as the OIC of security at Eakin Compound.

As we listened to General Eckhardt read the citation of award to Sergeant Froman, the transformation in Captain Giardino's entire countenance was so apparent that even as a casual observer, I could tell that something was troubling him greatly. After pinning the Silver Star medal on the swelling chest of Sergeant Froman and offering the usual congratulations and thanks for exemplary service, General Eckhardt directed his attention to Captain Giardino.

Having spoken but a few words, the General was interrupted by the Captain. Standing at attention, Captain Giardino barked out, "Sir!" This unexpected and extreme breach of military courtesy caused all eyes to rivet instantly upon the Captain. Breaking the thunderous silence that had enveloped us, Captain Giardino continued in a voice shaking with emotion, "I cannot in good conscience accept this award." This was strictly unheard of. He continued, "I have nothing against Sergeant Froman, and I add my congratulations to him. But I know exactly his duties over this past year, and I know exactly the duties I have been given as well. Honor forbids my acceptance of this far less distinguished recognition of service. Sir!"

All attention immediately focused upon General Eckhardt. I fully expected an explosion of outrage and indignation. I was to be disappointed in my expectation, but not in the General. General Eckhardt's expression was of deep understanding and concern. After a few thoughtful moments, he addressed Captain Giardino in a very soft and kind manner. "Captain, I want you to know that I personally appreciate your fine service, and I understand your disappointment. I hope that you will reconsider your decision and allow me the honor of presenting you with this medal."

The General's kindness and understanding took its toll on the Captain, and I watched as his frustration turned to sorrow. In sorrow he pleaded, "Sir. I'm sorry, Sir. I appreciate the opportunity you are giving me here, and I will always remember this with the greatest of personal respect for you, Sir. But I feel honor bound to decline this medal."

The General responded by shaking his head gravely and in the same conciliatory tone of voice said, "I'm sorry too, Captain. For now, you are dismissed."

After giving and receiving formal salutes, Captain Giardino immediately left through the door held open for him by the Sergeant Major.

The General turned his attention on me and with a wry smile on his face inquired, "Lieutenant Dinan, will you accept promotion to the rank of first lieutenant?"

I said, "Yes, Sir."

I was not well acquainted with Captain Giardino, and I had no knowledge concerning his military aspirations. Whatever they may have been, they ended right there, albeit with honor, dignity, and compassion. It is not a rule of thumb but rather a rule of rules: A captain does not confront his commanding general, especially in front of witnesses.

THIRTY-SIX

DON'T ASK

DURING this highly charged Tet period, an officer with the unlikely surname of Prevert managed to wend his way through the mists and report for duty at Eakin Compound. In a repeat of the old and venerated Christmas story, upon his arrival there was simply no room at the inn. We did not even have a stable. With not an unassigned bed to be found, the young private who was serving as charge of quarters that evening proved himself to be kind and caring by offering the travel-weary Captain the use of his not-to-be-used-that-night personal bunk for a night's slumber. The next morning Captain Prevert reported to Compound Headquarters where an empty desk eagerly awaited his occupancy. But there was still no bed for his sleep.

In an act of particular charity, the same young private offered the floor space between the bunks in his two-man quarters as a place for the Captain to sleep in this difficult time. Captain Prevert accepted the offer. Sometime in the small, dark, wee hours of that night, the young private awoke at the critical moment of what he supposed was a wet dream. Reaching for his member in the all-consuming darkness, he was shocked when he made hand contact with human flesh other than his own. He discovered quickly that he had put his hand on the face belonging to Captain Prevert. The why, as in why was the Captain's face so positioned, became immediately apparent. What to do or how to best handle this situation was altogether another matter.

There was no training time allocated, either in basic or in advanced individual training, to addressing the proper and acceptable military response for a soldier attacked in this fashion. Confused, embarrassed, but not stupid, the young man simply addressed the officer with a chilling, "Excuse me, Sir!" and rolled over onto his stomach, back to sleep. Rising with the sun, as Captain Prevert continued in his presumably contemptuous dreams, the confused and violated young soldier sought the counsel of the Compound First Sergeant.

Assessing the situation, the First Sergeant asked, "Do you want to press charges?"

The lad answered, "I don't care about charges. I simply don't want to see that man ever again."

Before the sun rose to the meridian, Captain Prevert had been reassigned to somewhere out in the boonies and whisked away from Eakin Compound with unmitigated dispatch. Nothing further ever came of his terrible response to the hospitality so kindly and generously given by the caring young private.

Learning of the situation, I was horrified by the incident yet bemused by the Captain's name as it related to his actions. It recalled to memory a case I had come across while reviewing dossiers as an enlisted man in the Intelligence Corps. At the time, I was assigned to the Department of Defense National Security Agency Check Center. In this capacity, along with a coterie of like-positioned young and inexperienced non-commissioned personnel, I was charged with reviewing multitudes of security portfolios.

Sharing any information gleaned from within these portfolios was strictly forbidden. Naturally, we shared with great glee anything we came across that might be in any way entertaining during our boring shifts. One such finding concerned a young WAC from the rural countryside of Kansas. Let's call her Judy.

Following the completion of her advanced individual training, this young private reported to her first duty posting at one of the many glamorous military installations located within the state of California. Her newly joined company was billeted in a classic World War II style barrack facility. The physical layout of this facility is familiar to all from the many movies depicting that period.

A large central hall with bunks lined up on two sides provided a communal sleeping area for all the resident privates. A separate room at one end of the hall was reserved for the company first sergeant, and the bathroom facilities were rather open in design, providing little, if any, privacy. The company commander, as a commissioned officer, had her own living quarters in an entirely separate building.

Shortly after settling in, Judy discovered that she definitely was not in Kansas anymore. This fact was punctuated by the relentless sexual advances made by most of her barrack mates. Like her corresponding male counterpart, Judy had received no training that might have prepared her to deal with this situation. And like her corresponding male counterpart, Judy sought the counsel of her first sergeant. At this point, we come to significant divergence in the story lines.

Assessing the situation, the first sergeant asked, "Do you think you might be more comfortable getting accustomed to having sex with other women if you were introduced to these pleasures by someone more senior and experienced, such as myself?"

This was clearly not the answer that Judy was seeking. Uncertain as to which way to turn, Judy took the risk of going over the sergeant's head by bringing her troubles to the attention of her company commander. To her chagrin, Judy was quickly to learn that the captain found her youth, innocence, and naiveté to be a temptation rather than a deterrent. So intrigued was this officer that she offered to break military regulations and establish a one-on-one relationship.

Private Judy returned to her barracks and wept. She did not know what to do. The unwanted and, to her, frightening advances continued nonstop. In desperation she went to the post's inspector general. The IG was not only a colonel but also a man. Having heard Judy's tale of woe, rather than initiate a hit on this young innocent, he initiated a full investigation of Judy's company. The investigation found that with the exception of Private Judy, every soldier, officer and enlisted alike, had a distinct preference for members of her own sex.

Judy was reassigned to another unit, and no charges were brought. Harmony in the ranks had been restored.

These incidences occurred in the 1960s, long before the "Don't ask, don't tell" formulation was invoked. Prior to the Stonewall Uprising, which established gay pride and brought our brothers and sisters inclined to this alternative lifestyle out of the closet, gays were, of course, in the military, and the military knew it. Whether in the civilian or military world, people simply did not ask and did not tell. As long as no one was truly harmed, most leaders in both worlds were able to elect the option of disinterested ignorance. When it became customary for military personnel to announce their preference and demand, "What are you going to do about it?" military leaders were deprived of this convenient and time-honored pretense and forced to take action. But what action should be taken?

"Don't ask, don't tell" is a formalization of a policy that functioned adequately when it, along with the affected population, operated clandestinely. Today, people tell without being asked. I have no idea where this is leading, except to note that new times cannot rely on old policy, especially now that the policy is written rather than whispered.

THIRTY-SEVEN

THE DECISION TO STAY

NOW that I'd been elevated to the exalted rank of first lieutenant, my life certifiably improved in the domains of compensation and respect. My augmented income allowed me to increase my savings account to the tune of not less than $500 a month. Remember inflation: in 1968, these US$500 were equivalent to an income increase of over US$5,000 in 2013! Also, as it is not easy to distinguish between a newly promoted first lieutenant and one on the verge of becoming a captain, an officer of this rank is accorded an element of respect that far exceeds the status enjoyed by second lieutenants.

February came to an uneventful close. March signaled the approaching end of my required 365-day tour of duty. Like almost every other member of the American military serving in Vietnam, I was required at this point to make a decision. Should I extend my tour of duty for six months and continue fencing with the devil I had come to know, or should I place my immediate destiny in the hands of Colonel Danforth or whichever factotum had replaced him? Inasmuch as the areas around Can Tho and Saigon had, for the most part, been completely pacified, I had no compelling concerns regarding my personal safety. This gave me the luxury of considering the pros and cons of my decision based on everything else.

If I determined not to extend my tour of duty in Vietnam, it was certain that my next assignment would be somewhere in the continental United States. To exactly where and in what capacity was unknowable. I had not made a decision as to whether or not I would extend my service beyond my required commitment to serve until the 19th day of January 1969 (two years following the date of my commission). If I opted for reassignment, I would be returned to America in the bloom of springtime April. But to where? The possibilities included so many hellholes against so few choice locations that the probability of a happy outcome on that score had long odds. I would be facing a minimum nine-month sentence wherever that might be. As to the nature of my workaday world in that unknown location, I was also unalterably in the dark.

The job I was doing and would continue to do at Eakin Compound was demanding, exhausting, and unrelenting. It was also challenging, exhilarating, and meaningful. I unquestionably had more responsibility and accountability than was usual for a lieutenant, and I enjoyed the personal satisfaction that comes with accomplishment. I did not relish the prospect of winding up in some dorky job where I would be, for all intents and purposes, a supernumerary. I was proud of what I was doing and convinced that how I did it benefited the lives of everyone touched by my operations. Further, I had power, and I liked being powerful.

What would be the consequences if I were to extend my tour in Vietnam for six months?

- I would be returned to America in the blast of late November. But with less than ninety days left on my commitment, I would be immediately separated from active duty should I decide not to extend my military career.
- I would continue to enjoy the distinction of serving as the Custodian of the Can Tho Mess Association until that time.
- I would continue to increase my saving account balance by a minimum of $500 every month.
- I would receive a thirty-day leave with paid roundtrip transportation to the destination of my choice.

- I would earn an additional seven day R&R vacation, including free transport.

I extended my tour of duty in Vietnam for six months.

Where I would take my leave was a no-brainer. I leapt at the opportunity to return to the U.S.A. with the specific and honorable intent of pursuing Anne Godwin Kinsey to the maximum extent possible. That maximum extent was defined by two distinct dates set in stone. Anne's graduation from Sweet Briar College was set for June 2nd, and on June 24th she was to sail out of New York harbor on the *SS France* to commence a summer-long European tour. I was told that due to final exams, the first day on which I would find a welcome on the Sweet Briar campus was the 30th of May. I determined not to arrive prior or subsequent to that date. I further determined that after June 24th, little allure would be left in my homeland.

Given this window of opportunity, I scheduled my departure for May 20th. I calculated that this would give me ample time for relaxation and reflection on both ends of my mission. This date was accepted by the requisite authorities, and orders were cut, sent, and received. In a state of high hopes and enthusiasm, I eagerly awaited my opportunity to endure an excruciating repeat of my inbound travel experience. Orders in hand, I posted letters to Anne and to my family in New York specifying an arrival date on the 22nd or 23rd of May. Like many a well-laid plan, this was to go awry.

THIRTY-EIGHT

SURPRISES

As it happened, one of my borrowed enlisted staff, Specialist Duke, had corresponding orders for deployment to the United States.

Duke had been born and raised in Las Vegas, Nevada. Eschewing the opportunity of attending an institute of post-secondary education, Mr. Duke migrated to Chicago and entered into the employ of Mr. Hugh Heffner. In this employment he fulfilled an unspecified but, as I was given to understand, much-needed interdisciplinary communication activity. Duke was not only smart, he was street-smart. The time he had spent cavorting with Mr. Heffner and his affiliates had increased his ability to manipulate less sophisticated operatives. Armed with this singular talent, Specialist Duke was one of the numerous enlisted personnel who, by dint of personal ability and experience, caused the military to attain the seemingly unattainable.

Specialist Duke had been assigned to the Army Corps of Engineers unit that operated out of Eakin Compound. Although he was well liked, his particular talents did not match up with the needs of this unit. When his commanding officer suggested that if he found a job more suitable to his abilities where the officer-in-charge was actually prepared to welcome him, he would release him to work for that officer, Specialist Duke wound up on my doorstep. When he arrived, my commissary operation was in its infancy, and since it was an unofficial operation, there was no one assigned to do the work. I desperately needed bodies and welcomed him in immediately. Sometimes you get lucky, and in this instance I got very lucky. Along with the need for personnel, we needed transport, not only in Saigon and in Can Tho, but between those two cities. Getting people moved about was often difficult but always doable. Moving large quantities of food after the conversion of all Vietnam operations to Class 1 messes was another matter altogether. Prior to that conversion, there had been transportation dedicated to provide for the specific need of moving supplies from the commissary in central Saigon to Tan Son Nhut airport and from there to various landing fields throughout the country. When the COLA system ended, that support ended also. This meant that in order to move supplies, we now had to scrounge, beg, and cajole for the use of needed trucks as well as air transport. This was Specialist Duke's specialty. Exactly how he did it, I never figured out. Some things are better accomplished by our enlisted personnel without officer interference. Somehow, whatever we needed to move got moved. Flight crews found space in their aircraft, and trucks materialized on demand. Without Duke, or someone very much like him, our commissary never could have been the success it was.

So it was that on May 20, 1968, Specialist Duke and I traveled together to Cam Ranh Bay and checked in at the air terminal. While we were checking in, we were diverted to the security office, where I was informed that there was a problem back at the mess association and that we could not leave Vietnam until it had been resolved. I called my office and spoke with Lieutenant King, who was filling in for me. The problem was a missing $400 worth of MPC. We acquired MPC from two sources: sales made in the clubs, snack bar, or commissary and MPC exchanged for Vietnamese piasters. The piaster exchange account was $60,000; the total sum value of piasters and MPC on hand should always equal that amount. Dealt with separately, and kept in a separate safe, were the MPC we collected from sales. The amount collected from sales was variable, and verification of the correct dollar amount came from our sales records.

A number of times we had experienced inconsistencies or differences between our physical count and the amount indicated by our sales records.

These inconsistencies had always been resolved when we found the record-keeping errors. We had never had an inconsistency in our piaster account. The missing $400 was from the piaster account. I assumed that there had been an inadvertent co-mingling of funds, and that the difference would be found in an error made in the sales records. I recommended rechecking this, that, and the other thing, which Lieutenant King graciously did, but in the end, no error was discovered, and Specialist Duke and I were instructed to return to Eakin Compound.

I was confident that I would be able to root out the clerical error responsible for this deplorable inconvenience. I verified the money count. I scoured every ledger and scrap of paper relevant to this abomination. I finally had to concede that the sum of 400MPC was indeed missing.

It wasn't very long after my arrival that my old nemesis, the Inspector General, Major Nordquest showed up. He arrived attempting to display a dignified, impartial professional concern. The joy that emanated not only from his face but from his entire person betrayed his real intent. After acknowledging his presence in the prescribed military manner, I succeeded in ignoring him completely.

After overtly observing every move made by each person in my office for about half an hour, the Major announced, "Lieutenant Dinan, I think you are acting very casually considering that you are a hair's breath away from being put under arrest." After I thanked him for his observation, he went on to inquire, "Do you have any idea what is going on here?" I told him that I had every idea possible and that I suspected that he had none. At this point, his fragile temper flared and he assured me, "If I were you, I'd be very careful. You're walking on very thin ice."

I hadn't planned to mouth off, but I was in an evil mood, and I harbored a distinct dislike for this particular smug malefactor. Without really thinking I said, "Major Nordquest, I appreciate the magnanimity of your personal visitation. If you have nothing positive to offer here, I suggest that you return to your office and initiate another one of your idiotic investigations. Like the others you have insisted upon, it will serve to prove my effectiveness while reacquainting your superiors with your willingness to waste government time and money to pursue a personal vendetta."

This unexpected response wiped the smile off his face and he left quickly with the admonition, "You will hear more from me on this matter."

Specialist Duke maintained a very low profile while my search was under way. When he learned that I had concluded that the money was unquestionably missing and that the next area of inquiry would focus on who might be responsible, he came forward and acknowledged that he would be on this list of suspects.

He explained, "You see, I relieved Private Pribble at the exchange window the day before we left, so I know that I will be on the top of the suspect list." While insisting upon his complete innocence, he went on to say, "With all the excitement of going home and all that, who knows? Maybe I did make a mistake."

I readily agreed that this was a possibility and asked, "What is your recommendation? I mean, what do you suggest that we do?"

"Well," Duke responded, "you know, it's really not my responsibility because I'm only an enlisted man, but I really want to get home. What if we split the difference and you and I both kick in $200? Would that settle the problem and get us out of here?"

"I don't know," I answered, "but I'll find out."

Something was definitely wrong, but the truth of the situation was secondary to me at that time. My most important goal was to get back to the United States, so I ran the proposal up through the chain of command. Happily, this simple solution met with approval, and after forking over $200, I was all set to go. But not so fast. New orders had to be cut, new airline tickets secured, and a new departure date specified. I was struck by the realization that my arrival in the United States would be delayed by a full week and that those expecting me would be worried out of their minds. Sending a letter was of no use. Mail between Can Tho and the states took longer than a week.

While I was engaged in my investigation, the gentleman in charge of the Red Cross office next door came in to fill his cup from my ever bountiful thermos pot. The gentleman in question had become accustomed to availing himself of our hospitality several times a day. Our relationship was cordial, professional, and friendly. We had never refused even one of his multiple, unending requests for free coffee service. We had never availed ourselves of any of the services under his direction. One of the services the Red Cross provided consisted of a few international phone lines setup at Can Tho Airfield. Using these lines, service personnel could actually place calls to America. The number of calls that could be put through each day was very limited, and access was controlled by a waiting list that stretched out for weeks. In emergency situations, that waiting list could be breached. To my way of thinking this was an emergency situation.

I took the Red Cross chief aside and explained my plight. He readily agreed that under the circumstances it would be horrible for my family back home to be left hanging in dread for seven, or possibly even more, days. He said that he would check with his staff at the phone call facility and set up a time slot for me. I was very appreciative and thanked him profusely.

He graciously said that I was over thanking him and that it was his pleasure. Then he went on, "Oh, just one more thing. I need to know who you will be calling. Will you be calling your parents or your wife?"

Had I known the rules, I would have lied without giving it a second thought. However, I was completely ignorant, so I made the mistake of telling the truth. I explained that I was orphaned at the age of fourteen, so calling my parents was not an option, and that I was still single, so I had no wife to call. Receiving this news, he became very grim and told me that since I had neither a parent nor a wife to call, I would not be allowed to use the Red Cross phone lines.

I was flabbergasted. "You must be joking."

"No, I'm not," he replied. "The rules are very clear. The phone lines are to be used for calling mothers, fathers, or wives and nobody else."

"This doesn't make any sense. What about brothers, sisters, grandmothers, or just family in general?"

"They don't count. The only calls allowed by Red Cross regulations are to wives and parents."

I tried to reason with him, but you can't very well reason with a regulation. He stuck to his regulation and my loved ones were stuck in fear. It seemed odd to me that when it came to getting help from the Red Cross, this organization would figuratively hang out a sign proclaiming: Help Available to Almost All; Orphans and Singles Need Not Apply. Something is fundamentally wrong with a charitable organization that can scorn a widow or an orphan.

I finally flew out of Vietnam on the 27th of May, and gaining a one day benefit by crossing the International Date Line, I arrived in New York on the 28th. Abbreviating my stay in New York, I hightailed myself to the Blue Ridge Mountains of Virginia and arrived at the Sweet Briar campus promptly on the 30th of May. The graduates and the grounds were more spectacular to behold than even my skewered memory anticipated. The graduation festivities and ceremony were in every way the epitome of understated, sophisticated glamour. Anne's Mom and Dad were not completely thrilled by my presence, but the General's daughter (Anne's grandfather was Brigadier General A. G. Strong) and Mr. Kinsey showed me many kindnesses in time-honored Southern style.

Following the festivities, Anne and I embarked on a whirlwind road trip, visiting people and places as suited our fancies, ending in New York City. Our delightful detour from the real world splashed to a finish with me standing lonely and dejected on a Hudson River pier waving valiantly as the *SS France* was towed away into deep waters.

Prior to escorting Anne to the boat, I had taken her to lunch at The "21" Club. We had been made to feel very welcome. I was both surprised and intrigued when Colonel Bob not only picked up our budget-busting bill but extended a military invitation: "Terry, I would like to visit with you before you return to Vietnam. Tomorrow at 10:30 will be convenient."

Although he was not my commander, he was possessed of that singular ability enjoyed by senior military officers to effect command. I made the only possible response, "Yes, Sir. I'll be here, Sir."

The next morning, just prior to the appointed time, I entered through the employee's entrance of this grand restaurant on Manhattan's fabled 52nd Street and was directed to the Colonel's sumptuous office. I knocked and received permission to enter. As I walked in, the Colonel made a theatrical observation of his wristwatch and indicated his satisfaction with the fact that I had presented myself at the exact time specified in his invitation. After inviting me to be seated, he instructed his secretary, Julia Dewitt, to invite Mr. Jerry to join us. Within moments Jerry Berns arrived, and Julie departed. After we exchanged a few pleasantries, Colonel "Mr. Bob" Kriendler inquired in a succinctly military fashion, "Lieutenant Dinan, have you considered your future plans?"

This was something I had been giving a great deal of thought to, and I had no trouble answering. I responded without hesitation, "Mr. Bob, the way I look at it, I have three major options at this time: I can stay in the Army and see what might happen there; I can accept the offer by the CIA to run their club in Saigon; or I can take advantage of the GI bill and return to college to earn my bachelor's degree."

Giving a nod that communicated respect for these suggested possibilities, the Colonel floored me by replying: "We would like to suggest a fourth option. We offer you the option of joining our management family here at '21.'"

I couldn't believe what I was hearing. No one unrelated to the Kriendlers or the Berns had ever been part of the management at The "21" Club, so this was definitely not something I had even dreamed of. I didn't have to tell Mr. Bob or Mr. Jerry that they had taken me by surprise. I knew that surprise was written all over me. I said simply, "Thank you. I accept. I'll be ready to go to work as soon as I return in mid-November."

The Colonel stood and, offering his hand, said, "Good, then it's all settled. We'll expect you to report for duty in November." The meeting was over.

I was in a state of shock. My future was planned. On top of everything else, I knew that Anne planned on moving to New York with three of her college friends as soon as she returned from Europe in September. Things were definitely looking up.

THIRTY-NINE

COMMENCEMENT

IT was already the beginning of July by the time I settled back in at Eakin Compound. With only four and a half months left to go, I soon developed a short-timer's perspective. I began thinking of all the details concerning signing everything over to my replacement and the job of preparing my things for shipment back to New York. So much to do, so little time, and I had to squeeze in a week's R&R vacation somewhere.

Days quickly became weeks and weeks seamlessly turned into months. In October, I spent a week of springtime in Australia, and when I returned to sunny South Vietnam I began counting at thirty days and a wake-up.

Evidently the search for my replacement developed into something of a quandary for those responsible. For the first time in a long time, they had to review the job description in order to match the job with exactly the right officer. Upon careful review, it was determined that one company grade officer could not be expected to adequately oversee an operation of this size and diversity. Somehow, unswayed by the fact that this one company grade officer had indeed done this for more than a year, it was determined that remedial organizational restructuring was required.

As a result, instead of dealing with one fellow officer, verifying inventories, employee records, monies, accounts, and equipment, I suffered the distinct honor of accommodating four bewildered replacements. Curiously to me, I was never asked to participate in the review of my operations and responsibilities. I was, however, pleased in an offbeat way when I learned that I, a lowly lieutenant, was to be replaced by four captains.

Captain number one was assigned to oversee the officer's club, senior NCO club, enlisted mens club, and the snack bar. Replacement captain number two was assigned to oversee the operation of the officers and enlisted mens mess halls. The third captain was assigned to operate the commissary, which at that time supported over 100 advisory teams in the field and was finally to be given official status. The fourth captain on my replacement roster was assigned to oversee the piaster exchange activity and the billeting services so dear to the hearts of the residents of Eakin Compound.

I found each of my replacements to be entirely professional and endlessly persnickety. Notwithstanding, in due course every ounce, inch, and stove under my control was signed off and registered as the responsibility of another officer. I was free at last. Free, free, *free* from the daunting responsibilities now assumed by my distinguished replacements.

Not surprisingly, my freely-offered advice to these replacements was freely declined. As senior officers, they each in their own way demonstrated umbrage at the prospect of receiving helpful instruction from a lower ranking officer. I, for one, could not have cared less. My only thought was *Get me out of here!*

I was airlifted out of Vietnam on the 14th of November. I arrived in New York City on the evening of the 15th and I reported for duty at 21 West 52nd Street on Monday, the 18th of November. It had been three and a half years since I had laid my spatula aside and removed my toque blanche for the final time before departing from that address and reporting to Fort Dix, New Jersey as an inductee for basic training in the United States Army. What a great adventure it had turned out to be. With that great adventure behind me, I welcomed the adventure now in front of me.

1969

GLOSSARY

AFB Air Force Base
APC Armored Personnel Carrier
AWOL Absent Without Official Leave
Billet/billeting Board and lodging; quarters for military personnel
Bird Dog The Cessna O-1 spotter plane (*see illustration*)
Cam Ranh Bay Cam Ranh Air Base, a USAF facility in Cam Ranh Bay, Vietnam
CIA Central Intelligence Agency
CG Commanding General
Charlie Army slang for Viet Cong/VC
CO Commanding Officer
Cobra The Bell AH-1 Cobra (*see illustration*)
Command Headquarters As used in this narrative, MACV IV Headquarters, located in downtown Can Tho
Compound Headquarters As used in this narrative, Eakin Compound Headquarters (*see diagram, building #21)*
COO Chief Operating Officer
COLA Cost of Living Allowance
CONEX Container Express bulk shipping containers
CORDS Civil Operations and Reactionary Development Support
CQ Charge of Quarters, an enlisted man assigned to the officer of the day; both are rotating assignments
Cyclo A three-wheeled motorcycle with two wheels up front and a seat for two in front of the driver; also spelled cycelo
DepCORDS The civilian diplomat appointed by the Commanding General MACV, to be his deputy in charge of CORDS
DMZ Demilitarized Zone
DoDNACC Department of Defense National Agency Check Center
E-1, E-2, et cetera Enlisted designations (*see U.S. Army Rank and Insignia, 1967–1968*)
Eagle's landing Pay day
ECOOM Eakin Compound Officers Open Mess (officers club)
EM Enlisted man/enlisted men
EM club, EC Enlisted mens club

EM mess Mess/dining room for enlisted men
Fahrenheit to Centigrade 120°F = 48.9°C; 110°F = 43.3°C; 100°F = 37.8°C
First Sergeant Special duty position held by the top E-8 enlisted person in a unit; often referred to as Top Sergeant or The Top
FLOOM Fort Lee Officers Open Mess (officers club)
Fourth Corps IV Corps, see below
Friendly Army slang for the anti-Communist Vietnamese who welcomed U.S. aid
G1 Department/Commanding Officer of Personnel
G2 Department/Commanding Officer of Military Intelligence
G3 Department/Commanding Officer of Operations and Training
G4 Department/Commanding Officer of Logistics (*see QMC*)
GI Anyone in the U.S. Army; from the abbreviation G.I. for Government Issue
Generals lounge A club facility exclusively for the use of the commanding general, the general's staff, and the general's specific invitees; aka the **general officers lounge**
Generals mess A mess exclusively for the use of the commanding general, the general's staff, and the general's specific invitees; aka the **general officers mess**
General staff G1, G2, G3, and G4
General's staff Officers chosen by the commanding general to assist him on a daily basis
IG Department of the Inspector General/the Inspector General
In-country In Vietnam
IV CORPS Fourth Corps, the southernmost military region encompassing the entire Mekong Delta; all U.S. Army operations in the Mekong Delta
LBJ Long Binh Jail, the American military jail in Vietnam
MACV Military Assistance Command Vietnam, headquartered in Saigon
MACV IV MACV IV Corps, all U.S. military operations in the Mekong Delta
Meshuggener Yiddish word meaning a crazy man
Mess Food services operation, mess operation, mess hall, dining room
Military time 24 hour clock (0001 hours–2400 hours)
MOS Military Occupational Specialty
MPC Military Payment Certificate/s, valued exactly the same as U.S. dollars
Naval rank/army rank equivalency
Naval captain is equivalent to army colonel
Naval lieutenant commander is equivalent to army major
Naval lieutenant is equivalent to army captain
NCO Non-commissioned officer
NCO club Non-commissioned officers club
NCOIC Non-commissioned Officer-in-Charge
O club, OC Officers Club
OCS Officer Candidate School
OD Officer of the Day, the officer who is on duty throughout the night
OER Officer's Efficiency Report

Officer grades
General grade: generals and colonels
Field grade: lieutenant colonels and majors
Company grade: captains and lieutenants

OIC Officer-in-Charge
OM Mess/dining room for officers
Piaster Monetary currency of the Republic of Vietnam
Piaster exchange facility Post facility for the exchange of MPC into piasters
PSYOPS Army Psychological Operations, also written as PsyOps
Puff the Magic Dragon The Douglas AC-47, also known as Spooky (*see illustration*)
PX Post Exchange, an all-purpose store
QM Quartermaster
QMC Quartermaster Corps, the logistics branch of the U.S. Army; it provides billeting, food services, laundry, currency exchange services, et cetera
R&R Rest & Recuperation leave
Recon Reconnaissance
Re-up Extend one's tour of duty, which was twelve months in Vietnam
ROTC Reserve Officers' Training Corps, a program for college students
Roger OK, it's a go
Round-eye Non-Asian
Senior NCO Club Non-commissioned officers club for senior NCOs, E-6 and up
SNAFU Military jargon: Situation Normal All Fucked Up
TEAM #96 MACV Advisory Team #96—the official designation of the unit to which all army personnel at Eakin Compound were assigned or attached
Tan Son Nhut Tan Son Nhut Air Base, on the outskirts of Saigon
Tet Tet Nguyen Dan, Vietnamese New Year, often used in reference to the 1968 Tet Offensive
TOE Table of Organization and Equipment
Top Sergeant see First Sergeant
Top Secret Highest military clearance; the sequence is Top Secret, Secret, Confidential
Under arms Carrying weapons
Unfriendly Army slang for the enemy, Communist Vietnamese fighters
USAF United States Air Force
USAID United States Agency for International Development
USARV United States Army Republic of Vietnam
USMC United States Marine Corps
USO United Service Organizations, a non-profit civilian entity that supports and entertains U.S. troops worldwide
VC Viet Cong, Vietnamese Communist guerilla fighters
WAC Women's Army Corps

VIETNAMESE WORDS USED IN THIS TEXT

Ao dai Traditional Vietnamese silk dress with a tightly fitted top, long sleeves and flowing skirt panels over loose pants
Ba-muoi-lam Thirty-five
Choi oi General exclamation along the lines of Oh my word, Holy Cow, What the hell!
Co Miss, female
Dai Uy Captain
Mon joi Hello, Good morning
Ong Mister, male
Thieu Uy Lieutenant

U.S. ARMY RANK and INSIGNIA, 1967–1968

Insignia immediately identifies military rank. Each increase in rank represents growing leadership ability and increased responsibility.

HIERARCHY

Rank	Abbreviation
OFFICERS	
General of the Army	*none*
General	*none*
Lieutenant General	Lt. Gen.
Major General	Maj. Gen.
Brigadier General	Brig. Gen.
Colonel	Col.
Lieutenant Colonel	Lt. Col.
Major	Maj.
Captain	Cpt.
First Lieutenant	1st Lt.
Second Lieutenant	2nd Lt.
WARRANT OFFICERS	
Chief Warrant Officer (4)	CWO
Chief Warrant Officer (3)	CWO
Chief Warrant Officer (2)	CWO
Warrant Officer	WO

NON-COMMISSIONED OFFICERS

Sergeant Major of the Army (E-9)	SMA
Command Sergeant Major (E-9)	Cmd. Sgt. Maj.
Sergeant Major (E-9)	Sgt. Maj.
First Sergeant (E-8)	1st Sgt.
Master Sergeant (E-8)	M. Sgt.
Sergeant First Class (E-7)	Sfc.
Staff Sergeant (E-6)	S. Sgt.
Sergeant (E-5)	Sgt.
Corporal (E-4)	Cpl.

SPECIALISTS

Specialist Seven (E-7)	Spec. 7
Specialist Six (E-6)	Spec. 6
Specialist Five (E-5)	Spec. 5
Specialist Four (E-4)	Spec. 4

OTHER GRADES

Private First Class (E-3)	Pfc.
Private (E-2)	Pvt.
New Recruit (E-1)	Pvt.

INSIGNIA

COMMISSIONED OFFICERS
Shoulder Insignia

General Grade

GENERAL OF THE ARMY
(Appointed only in wartime)

GENERAL

LIEUTENANT GENERAL

MAJOR GENERAL

BRIGADIER GENERAL

COLONEL

Field Grade

LIEUTENANT COLONEL

MAJOR

Company Grade

CAPTAIN

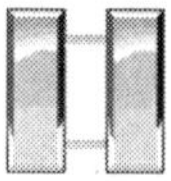

FIRST LIEUTENANT

SECOND LIEUTENANT

All commissioned officer insignia are silver in color except the Major's gold oak leaf and the Second Lieutenant's gold bar.

WARRANT OFFICERS
Shoulder Insignia

CHIEF WARRANT OFFICER 4

CHIEF WARRANT OFFICER 3

CHIEF WARRANT OFFICER 2

WARRANT OFFICER

CWO4 and CWO3 insignia are brown with silver stripes.
CWO2 and WO insignia are brown with gold stripes.

NON-COMMISSIONED OFFICERS
Sleeve Insignia

SERGEANT MAJOR OF THE ARMY,
COMMAND SERGEANT MAJOR (E-9)

SERGEANT MAJOR (E-9)

FIRST SERGEANT (E-8)

MASTER SERGEANT (E-8)

SERGEANT FIRST CLASS (E-7)

STAFF SERGEANT (E-6)

SERGEANT (E-5)

CORPORAL (E-4)

SPECIALISTS
Sleeve Insignia

SPECIALIST SEVEN (E-7)

SPECIALIST SIX (E-6)

SPECIALIST FIVE (E-5)

SPECIALIST FOUR (E-4)

OTHER
Sleeve Insignia

PRIVATE FIRST CLASS (E-3)

UNITED STATES ARMY QUARTERMASTER CORPS

Branch Insignia: A gold eagle with wings outspread sits on a gold wheel with thirteen spokes, a blue rim decorated with thirteen stars, and a red and white hub. A gold sword and key are crossed over the wheel. The insignia is three-quarters of an inch high.

The wagon wheel symbolizes transportation; the thirteen spokes and stars represent the thirteen colonies and the inception of the Quartermaster Corps in the Revolutionary War. The sword and the key to the storeroom indicate its control of military supplies. The eagle, representing freedom, is the national emblem of the United States.

Branch Plaque: The plaque displays the branch insignia on a buff ground encircled by a blue band, edged and lettered in gold.

MILITARY ASSISTANCE COMMAND VIETNAM (MACV)

MACV patch: The MACV shoulder sleeve insignia is a red shield bordered in gold with a white sword standing on its gold hilt.

CPSIA information can be obtained at www.ICGtesting.com
Printed in the USA
BVOW02s2321180815

413929BV00003B/3/P